Introduction to Inorganic Chemistry

Introduction to Inorganic Chemistry

Dennis Close

NY Research Press
New York

Published by NY Research Press
118-35 Queens Blvd., Suite 400,
Forest Hills, NY 11375, USA
www.nyresearchpress.com

Introduction to Inorganic Chemistry
Dennis Close

International Standard Book Number: 978-1-63238-813-1 (Hardback)

Cataloging-in-Publication Data

Introduction to inorganic chemistry / Dennis Close.
p. cm.
Includes bibliographical references and index.
ISBN 978-1-63238-813-1
1. Chemistry, Inorganic. 2. Chemistry. 3. Inorganic compounds. I. Close, Dennis.
QD151.3 .I58 2020
546--dc23

Printed in China.

Contents

Preface

The chemical compounds which lack carbon-hydrogen bond are known as inorganic compounds. Inorganic chemistry is a branch of chemistry that focuses on the study of the behavior and synthesis of inorganic compounds. Inorganic chemistry is broadly divided into a few major sub-fields which are involved in studying different aspects of inorganic compounds. Some of these sub-fields are descriptive inorganic chemistry, theoretical inorganic chemistry and mechanistic inorganic chemistry. It is utilized in diverse industries such as materials science, surfactants, medications, fuels, pigments and agriculture. This book is a valuable compilation of topics, ranging from the basic to the most complex theories and principles in the field of inorganic chemistry. Some of the diverse topics covered herein address the varied branches that fall under this category. For all those who are interested in inorganic chemistry, this textbook can prove to be an essential guide.

To facilitate a deeper understanding of the contents of this book a short introduction of every chapter is written below:

Chapter 1- Chemistry is broadly divided into two sub-disciplines, inorganic chemistry and organic chemistry. The branch of chemistry which deals with the behavior and synthesis of inorganic and organometallic compounds is known as inorganic chemistry. This is an introductory chapter which will introduce briefly all the significant aspects of inorganic chemistry.

Chapter 2- A solid can either exhibit crystalline structure or amorphous structure. The most basic repeating unit in a crystal is known as a unit cell. The bonding of atoms is explained using a variety of different theories such as valance bond theory and molecular orbital theory. The chapter closely examines these key concepts of structure and bonding in inorganic chemistry to provide an extensive understanding of the subject.

Chapter 3- A compound which does not have a carbon to hydrogen bond is known as an inorganic compound. Some of the common inorganic compounds are water, halogens and halides and metal hydrides. The topics elaborated in this chapter will help in gaining a better perspective about the properties of these inorganic compounds.

Chapter 4- A compound which contains coordinate bonds is known as a coordination compound. Some of the different properties exhibited by coordination compounds are denticity, hapticity and chirality. This chapter has been carefully written to provide an easy understanding of these varied facets of coordination compounds.

Chapter 5- Reactions in inorganic chemistry are divided into a few broad categories, namely, combination reactions, decomposition reactions, replacement reactions and combustion reactions. The chapter closely examines these key reactions in inorganic chemistry to provide an extensive understanding of the subject.

Chapter 6- Elements within the periodic table are arranged on the basis of their properties into different groups. A few of these groups are alkali metals, alkaline earth metals, boron group and carbon family. The topics elaborated in this chapter will help in gaining a better perspective about the different groups of elements.

I owe the completion of this book to the never-ending support of my family, who supported me throughout the project.

Dennis Close

1

Inorganic Chemistry: Basics

Chemistry is broadly divided into two sub-disciplines, inorganic chemistry and organic chemistry. The branch of chemistry which deals with the behavior and synthesis of inorganic and organometallic compounds is known as inorganic chemistry. This is an introductory chapter which will introduce briefly all the significant aspects of inorganic chemistry.

Chemistry

Chemistry is a physical science concerned with the composition, structure, behavior, and properties of matter and with the changes it undergoes during, and as a result of, chemical reactions. It involves study of substances in all of the states of matter (solid, liquid and gas) and knowledge and understanding of the various structures of matter (including e.g. atoms, molecules, crystals and other aggregates) whether in isolation or in combination with others.

Areas of Chemistry

The study of modern chemistry has many branches, but can generally be broken down into five main disciplines, or areas of study:

- Physical chemistry: Physical chemistry is the study of macroscopic properties, atomic properties, and phenomena in chemical systems. A physical chemist may study such things as the rates of chemical reactions, the energy transfers that occur in reactions, or the physical structure of materials at the molecular level.
- Organic chemistry: Organic chemistry is the study of chemicals containing carbon with hydrogen. Carbon is one of the most abundant elements on Earth and is capable of forming a tremendously vast number of chemicals (over twenty million so far). Most of the chemicals found in all living organisms are based on carbon.
- Inorganic chemistry: Inorganic chemistry is the study of chemicals that do not, in general, contain carbon. Inorganic chemicals are commonly found in rocks and minerals. One current important area of inorganic chemistry deals with the design and properties of materials involved in energy and information technology.
- Analytical chemistry: Analytical chemistry is the study of the composition of matter. It focuses on separating, identifying, and quantifying chemicals in samples of matter. An analytical chemist may use complex instruments to analyze an unknown material in order to determine its various components.
- Biochemistry: Biochemistry is the study of chemical processes that occur in living things.

Inorganic Chemistry

Inorganic chemistry is one of the Branches of Chemistry, involves all element as the study domain of organic chemistry meant to be study of hydrocarbon then inorganic chemistry study everything else. This statement means inorganic chemistry has wider study because it involves all of element written on periodic table as well as carbon who has significant role in inorganic chemistry.

The organometallic field bridge these two areas of study since it compounds containing metal-carbon bonds plus it is also includes catalysis of many organic reaction. Bioinorganic chemistry bridges biochemistry with inorganic chemistry and has prominent position on medical application. In short, inorganic study is a vast subject with limitless areas for investigation and potential practical application.

Technically, the difference between organic and inorganic chemistry is on its coordination number, while organic chemistry has maximum coordination number of 4 inorganic chemistry has coordination number exceed of 4. Inorganic chemist usually solves problems to fix the properties, structures, and reactivity's materials. So, they must concern with a great method of synthesis, manipulation, and characterization of inorganic compounds.

Most of inorganic compound are formed of ionic compounds consist of cations and anions joined by ionic bonding. Some of classes in inorganic compounds are oxides, the carbonates, the sulfates, and halides most of them could be characterized by its high melting point.

Nuclear Chemistry

An atom consists of a positively charged nucleus who has extra-nuclear electrons around it. The effect of nucleus despite of the fact that the mass of atom is almost entirely concentrated in it. On chemical behavior is exerted through its influence on the number, and hence the arrangement of the electrons.

It can be argued that discussion of nucleus or nuclear is more appropriate as physic but since changes in the nucleus in an atom must be followed with chemical behavior. Also, as it's because of the increasing interpenetration of physic and chemistry the study of nuclear transformation is the domain for chemistry study.

Nuclear changes can and occur in unknown time as it could be happened spontaneously, as the phenomena of radioactivity which is the main deal of nuclear chemistry along with nuclear processes. Nuclear transmutation and nuclear properties but some of it can be brought about experimentally in the laboratory. These changes can and will be classified as (i) natural radioactivity; (ii) changes caused by the bombardment of nuclei; (iii) nuclear fission and (iv) nuclear fusion.

Natural Radioactivity

Radioactivity is now to be seen as a property of the nucleus and as consequence of the inherent stability of some nuclei. Thus, the nuclei of the atoms of all naturally occurring element whose atomic members are greater than 82 are unstable and so undergo slow decay. These elements are said to be radioactive, to exhibit radioactivity and to undergo radioactive disintegration or decay.

The natural radioactivity of elements has been used in practical world such as medicine in the treatment of cancer and other malignant growth. For the manufacture of luminos enamels and as one means of calculating the age of the earth.

The elements such as radium and radon are used for the treatment of cancerous growth. The process start with the radiations from these elements who are capable to destroy the cancerous cell on the human tissue. Thus, the side effect they can also destroy healthy tissue so it is needed to protect the healthy areas when the treatment is on progress by shielding it.

Transmutation by Nuclear Bombardment

The transmutation of elements is applied on base metals into gold, who has fascinated mankind for many centuries which clearly because of the operation of the alchemist. As we know, the changes by radio elements like transmutation is taking place naturally but so far there is no method found to altering. Even in slightest degree, the rate at which these changes are going on until these recent years.

The transmutation is mainly about the conversion of one chemical element or an isotope into another. Because istope is defined by its proton number, nuclear transmutation occurs in any process that make changes in proton or neutron number.

Nuclear Fission

Firstly, found by German Otto Hahn and his assistant Fritz Strassmann in 1938 nuclear fission later explained theoretically by Lise Meitner and her nephew Otto Robert Frisch a year later. Fission is a form of nuclear transmutation which is happened because of the process of splitting an atom into smaller part called nuclei, which generating free neutrons and gamma photons other than generating neutrons and gamma photon. It also releases very large amount of energy by energetic standards or radioactive decay.

Typically, the two nuclei produced are comparable and have different sizes. The energy released contained in nuclear fuel is times the amount of free energy contained in a similar mass of chemical fuel such as gasoline, making nuclear fission a very dense source of energy.

Nuclear Fusion

In principle, nuclei is very light and it can be combined to form a heavier ones and release bigger energy as well as the light one. This process called fusion the contrast of fission but also resulting energy because of the difference in mass arises due to the difference in atomic binding energy between atomic nuclei before and after reaction. Such process is the main source of the energy generated in the sun and the other stars. These process also form the basis of the hydrogen bomb.

Today, research is focusing on way to adapt nuclear fusion process to the controlled, sustained generation of energy but the practical result is still far from now. The reason that fission and fusion are sources of nuclear energy can be understood by referring to a plot of the binding energy per nucleon as function of mass number.

Thermochemistry

Thermochemistry is the branches of inorganic chemistry. This field concerns on chemical reactivity.

Chemical Energy

The study of the heat changes followed by chemical action is often referred to as the study of Thermochemistry. All chemical substances or element are associated with intrinsic energy, which mostly in the form of heat which go less or higher degree when it undergoes a chemical reaction.

Any chemical system either single substance or consist of group of substances contains an amount of energy this amount are influenced by the mass of its substance, chemical nature and physical condition of the system.

This condition will remain constant until a reaction takes place. Thus, the production of the reaction will constitute a chemical system of different intrinsic energy. Even though, we could not know the absolute value of a total energies stored in intrinsic energy, we could determine the change in total energy when the initial system passes into a new system and if there was no external work has been done the decrease in intrinsic energy will be equal to the heat evolved, or vice versa. Endothermic reaction occurs when system absorb heat whilst exothermic is a term for reaction when system releasing heat.

Heat of Reaction

Experiment shows that the heat effect with any given chemical changes has a constant value for a given quantity of reactants. When this value is expressed as the quantity of heat liberated or absorbed during the reaction of the gram-molecular quantities of the reacting substances as indicated by the equation for the reaction known as the Heat of Reaction. The amount of heat is usually represented in Kilograms-calories.

The heat of reaction will depend on physical state of reacting substances and of the products of reaction. Thus, this ought to be indicated when giving the value of heat reaction. The quantity of heat liberated or absorbed when one gram-molecule of a compound is formed from its element called the Heat of Formation of compound, whilst the quantity of heat liberated when one gram-molecule of an element or compound is completely oxidized is called the Heat of Combustion.

Hess's Law

In 1840, G.M. Hess measured the heat developed during the formation of a compound by several experiment and come to conclusion that "the amount of heat evolved during the formation of a given compound is the same whether the compound is formed directly all at once or slowly."

This means if a chemical change in different routes the overall enthalpy change remain the same. Hess's law can be used to determine the overall energy required for a chemical reaction. When it can be divided into synthetic steps so it will be easier to characterize. This may be used as a basis to design complex syntheses.

Thermochemistry Process

A process happened if one or more properties of a system changes. Process could be divided into three categories; Isobaric means the process occurs when the pressure of system remaining constant; Isothermal means the process occurs when the temperature of system remaining constant; Adiabatic means process occurs without any heat exchange.

Theoretical Chemistry

Theoretical chemistry is a study focusing on explanation of chemical and physical observation of atom, this study includes physic law as its base such as Columb's law which describes force interacting between static electrically charged particles, Kinetics energy which explain that an object has energy because of it motion, virial theorem and so on other laws.

Appropriate level of theory is needed to explain an observation thus some theoretical method has to be match its own observation or in other word the right theoretical method or approach is needed to explain an observation to decrease the bias degree of the theory made.

Quantum Theory

One of the most interesting and also one of most important problems in the early development of chemistry is the nature of radiant energy. In eighteenth century, most of scientist accepted the idea that visible light consisted of small particles that were emitted from the source like bullets. Such theory was propose by Sir Isaac Newton in Royal Society in 1675, and most of universe accepted his views.

Moreover, quantum chemistry has been focused on the application of quantum mechanic in physical models and experiments of chemical system, which commonly also known as molecular quantum mechanic because of its strong relation with molecule.

Computational Chemistry and Mathematical Chemistry

Either computational or mathematical, both of them are tools to re-model structure of molecule by terms of prediction. These method used as alternative so it is not necessarily referring to quantum mechanic. In mathematics, topology can be used to predict properties of flexible finite size bodies like clusters.

The application of computer codes allows us to predict the molecular shape properly by using some method such as Hartree–Fock, post-Hartree–Fock, density functional theory, semi empirical methods (such as PM3) or force field methods. Besides of predicting the shape of molecule properly computer is able to predict vibrational spectra and vibronic coupling, also acquire and transform Infra-red data into frequency information. The use of computer and informational techniques, applied to crop information to solve problems in the field of chemistry also called Cheminformatics.

Molecular Dynamic and Molecular Mechanic

In molecular dynamic, we apply the classical mechanic to simulate the movement of nuclei (smaller part of nuclear) in its process of assembling of atoms and molecules. Van der Waals forces controlled the rearrangement of molecules and promoted by temperature.

If molecular dynamic apply the classical mechanic to simulate the movement of nuclei, molecular mechanic tries to use classical mechanic to model the molecular system. So, it could be easier to calculated the potential energy stored in molecular system by using force fields.

Moreover, the method used to modelling the structure of atoms called Molecular Modelling such as molecular docking, protein-protein docking, drug design, combinatorial chemistry.

Photochemistry

Photochemistry is a branch of chemistry that domain of study involves chemical reaction caused by the absorption of ultraviolet. When a molecule absorb a photon of light, this changes the electronic structure, and it reacts differently with other molecules. The result of the energy absorbed from light are photochemical changes in the absorbing molecule, or in an adjacent molecule for example photosintization.

The energy also can be formed as heat, or as lower energy light, for example fluorescence or phosphorescence, in order to return the molecule to its ground state. Each type of molecule has a different preference for which of these different mechanisms it uses to get rid of absorbed photon energy, for example some molecule because of it type prefer fluorescence than chemistry.

Photochemistry Applications

There several photochemistry reaction because of each type of molecule prefer different ways to release the energy they have been absorbed, some of application example of photochemistry reaction are:

- Photosythesis: The process where plant uses ultraviolet energy to convert dioxide and water into glucose and oxygen.
- Bioluminesence: Fireflies produce enzym in their belly to generate light.
- Photodynamic Theraphy: The use of light to destroy tumor on human body by the action of singlet oxygen generated by photosensitized reactions of triplet oxygen. Typical photosensitizers include tetraphenylporphyrin and methylene blue.
- Photoresist: Is a light-sensitive material used in several processes to form a patterned coating on a surface which has significant role in the whole electronic industry.

Photochemistry Law

Laws in photochemistry can be used and developed in the study of photochemistry there are a lot of rules that applied for this area of study but some of them are:

- The First Law of Photochemistry states that light must be absorbed for photochemistry to occur. This is a simple concept, but it is the basis for performing photochemical and photobiological experiments correctly.
- The Second Law of Photochemistry states that for each photon of light absorbed by a chemical system, only one molecule is activated for a photochemical reaction.
- The Bunsen-Roscoe Law of Reciprocity states that a photochemical effect is directly proportional to the total energy dose, irrespective of the time required to deliver the dose.

These laws has affected the study in photochemistry and could be used in certain situation as these term and condition applied for each of this rule.

Geochemistry

Geochemistry is a study that bridges earth study which the focus is understanding earth system with chemistry, in other word to put this explanation of geochemistry is geochemistry use chemistry as it approach to deepen understanding of how earth system works. Sub-disciplines of geochemistry include biogeochemistry, organic geochemistry, trace and elemental geochemistry, and metamorphic and igneous-rock geochemistry.

In 1838, a Swiss-German chemist Christian Friedrich Schönbein was the first person who use term of geochemistry in his paper by stating a prediction that a new field of study will be born "In a word, a comparative geochemistry ought to be launched, before geochemistry can become geology, and before the mystery of the genesis of our planets and their inorganic matter may be revealed".

- Biogeochemistry: There are sub-fields of geochemistry such as biogeochemistry which focusing on the effect of life on chemistry point of view on earth.
- Organic Geochemistry: Organic geochemistry who involves the study of role processes and derived compounds from living or once-living organisms.
- Trace Elemental Geochemistry: The trace elemental geochemistry which focusing on understanding the origin and evolution of rocks.
- Metamorphic Igneous-Rock Geochemistry: Thus, metamorphic igneous-rock geochemistry which the main field study is the transformation of rock in chemistry point of view.

Solid State Chemistry

Solid state chemistry usually refer to term of material chemistry, a study about synthesis, structure, and properties of solid phase materials, particularly, but not necessarily exclusively of, non-molecular solids.

Solid State Chemistry Fields

Moreover, solid state chemistry is mainly about understanding of the processes that occur during formation and growth of crystals and its application into the prediction and control of crystal morphology and perfection.

The field of solid state inorganic chemistry is very wide area with enermous practices. There are a lot of method involves in synthesizing of state compound some of them are oven techniques, melt methods, solution methods, gas reaction and air & moisture sensitive materials. Each of this technique differs because of differentiation of character of solid state compounds.

Meanwhile, there six sub-branches of Inorganic Chemistry. These knowledge is important in producing new compound that could be useful in life. Indeed, chemistry held the basic compound to create new invention whether in industry field, technology, or space.

Physical Techniques in Inorganic Chemistry

Spectroscopic and Structural Methods

1. UV-Vis Absorption Spectroscopy: The number, energies and intensities of a transition metal compound's absorption bands in the UV-Vis and near IR can be used to determine the general type of atom bound to a metal and the geometry about the metal.

2. Circular Dichroism (CD) Spectroscopy: CD spectroscopy measures the degree to which a sample rotates circularly polarized light. CD is limited to enantiomerically pure, optically-active compounds; an equal mixture of enantiomers will give no CD spectrum (enantiomeric pairs having CD spectra that are mirror images of each other and so would cancel out). CD is similar to absorption spectroscopy, but because of different selection rules for the electronic transitions, peak intensities and widths may be different between CD and absorption (the energy of a peak must still, however, be the same). This is a very useful technique for studying metal-containing proteins and enzymes because the chirality of the peptide induces a CD signal from any metal bound to the protein.

3. IR Absorption Spectroscopy: Can be used just like in organic chemistry to fingerprint a compound. In simple compounds the number, energy and intensity of the IR transitions are directly related to the geometry of compound and to which atoms are bound to which other atoms. Unfortunately, some metal-ligand vibrations, and many of the vibrations for the heavier elements occur outside the frequency window of most commercial instruments. For complex compounds involving large organic moieties, the IR becomes more difficult to interpret. IR is useful to determine the presence of complex counter ions like PF_6^-, ClO_4^-, BF_4^-, because they have distinctive absorptions in the IR.

4. Raman Spectroscopy: Raman spectroscopy is complementary to IR absorption spectroscopy. Both probe vibrations within a compound, but they have different selection rules. By considering what peaks are present, or absent, in the two spectra of a compound, one can determine geometry; at least in simple cases. Raman is also useful because it can, depending on instrument design, scan to very low frequencies (~100 cm^{-1}), and thus observe transitions too low for IR absorption. A variant technique, resonance Raman, can be used to assign vibrations and identify ligands.

5. Mass Spectroscopy: In inorganic chemistry this is most often used to determine the molar mass of compounds. When mass spectroscopy data are combined with an elemental analysis, the chemical formula of the substance can be determined. Analysis of a compound's fragmentation pattern can be used to gain structural information. This is not usually done because of the complex fragmentation patterns of inorganic compounds, and because other methods are available for structure determination.

6. Nuclear Magnetic Resonance (NMR): The workhorse of chemistry. While paramagnetic NMR is not impossible, NMR is usually performed on diamagnetic compounds. The NMR spectra of inorganic compounds are often more complicated than organics because other nuclei also have nuclear magnetic moments. So in addition to the familiar ^{1}H (I = 1/2) and ^{13}C (I = 1/2), there are ~90 other elements that have at least one NMR-active nucleus. Although less widespread than the standard solution NMR, solid state NMR and even single-crystal NMR have been used on materials that simply do not dissolve in any solvent.

7. Magnetism: The number of unpaired electrons in a transition metal compound is a very useful physical property to know. From a knowledge of the magnetism one can determine what metal is present, its oxidation state and even a rough idea of the metal's structure. Magnetic measurements are often made with a Gouy balance, an Evans balance (a modified version of the Gouy balance), or a SQUID (super-quantum interference device) magnetometer. A compound's magnetic moment can also be determined by NMR using the Evans method. Its one drawback is that it is a bulk technique, and when used on mixtures the measured magnetic moment will be the weighted average of the moments of all of the paramagnetic species present.

8. Electron Paramagnetic Resonance (EPR or ESR): While NMR is usually only for diamagnetic compounds, EPR is for paramagnetic compounds with an odd number of unpaired electrons (EPR can be done when there are an even number of unpaired electrons, but it is a much harder experiment). In this experiment a sample is irradiated with microwave radiation and the field is swept until resonance occurs. The field at which resonance occurs depends on the number of unpaired electrons, the geometry about the metal center and the metal's ligands. In many ways EPR and NMR the same, and there is even a technique that combines them (ENDOR, electron-nuclear double resonance, spectroscopy).

9. Magnetic Circular Dichroism (MCD): MCD is a hybrid technique based on the fact that all matter will rotate circularly polarized light in the presence of a magnetic field. It is most useful for paramagnetic compounds (either with an even or odd number of unpaired electrons). Because it is both a magnetic and a spectroscopic method, MCD can be used to measure a compound's magnetic properties and electronic transitions. MCD is most powerful when used in conjunction with another method. It can also be used on a mixture, if transitions arising from different species can be identified. The main limitations of MCD is that it is usually performed at low temperature ($<$ 77 K) and the samples must be strain-free glasses. The cost of the instrumentation and the cryogens have limited MCD to only a few groups worldwide.

10. X-Ray Methods: Single-crystal X-ray diffraction is the most powerful X-ray technique for inorganic chemists. From precise measurement of the intensity and angles at which an X-ray beam diffracts off a crystal, the arrangement of the atoms can be reconstructed. Obviously, as a direct probe of structure crystallography is an invaluable characterization method for all types of compounds. Some inorganic compounds (e. g., rocks, minerals) can't be obtained as single crystals. In these cases X-ray powder diffraction can be used to obtain the dimensions of the unit cell for use in identification (there is a large, indexed catalog of lattice constants for many minerals).

2

Inorganic Chemistry: Structure and Bonding

A solid can either exhibit crystalline structure or amorphous structure. The most basic repeating unit in a crystal is known as a unit cell. The bonding of atoms is explained using a variety of different theories such as valance bond theory and molecular orbital theory. The chapter closely examines these key concepts of structure and bonding in inorganic chemistry to provide an extensive understanding of the subject.

Unit Cells and Crystal Structures

The structure of solids can be described as if they were three-dimensional analogs of a piece of wallpaper. Wallpaper has a regular repeating design that extends from one edge to the other. Crystals have a similar repeating design, but in this case the design extends in three dimensions from one edge of the solid to the other.

We can unambiguously describe a piece of wallpaper by specifying the size, shape, and contents of the simplest repeating unit in the design. We can describe a three-dimensional crystal by specifying the size, shape, and contents of the simplest repeating unit and the way these repeating units stack to form the crystal.

The simplest repeating unit in a crystal is called a unit cell. Each unit cell is defined in terms of lattice points the points in space about which the particles are free to vibrate in a crystal.

The structures of the unit cell for a variety of salts are shown below.

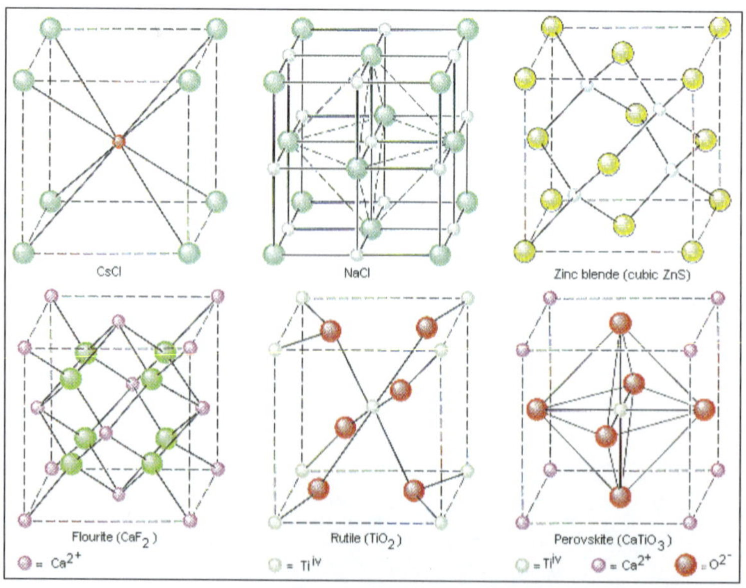

In 1850, Auguste Bravais showed that crystals could be divided into 14 unit cells, which meet the following criteria:

- The unit cell is the simplest repeating unit in the crystal.
- Opposite faces of a unit cell are parallel.
- The edge of the unit cell connects equivalent points.

The 14 Bravais unit cells are shown in the figure below.

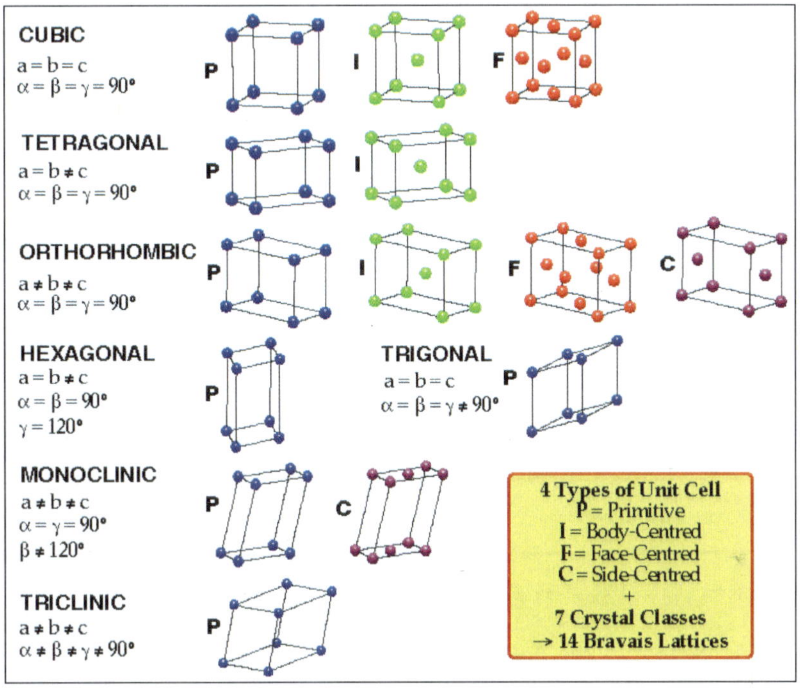

Cubic category, which includes the three types of unit cells--simple cubic, body-centered cubic, and face-centered cubic-shown in the figure below.

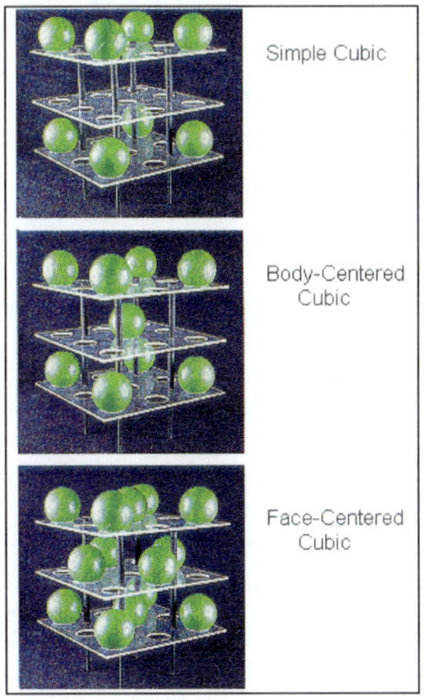

These unit cells are important for two reasons. First, a number of metals, ionic solids, and intermetallic compounds crystallize in cubic unit cells. Second, it is relatively easy to do calculations with these unit cells because the cell-edge lengths are all the same and the cell angles are all 90.

The simple cubic unit cell is the simplest repeating unit in a simple cubic structure. Each corner of the unit cell is defined by a lattice point at which an atom, ion, or molecule can be found in the crystal. By convention, the edge of a unit cell always connects equivalent points. Each of the eight corners of the unit cell therefore must contain an identical particle. Other particles can be present on the edges or faces of the unit cell, or within the body of the unit cell. But the minimum that must be present for the unit cell to be classified as simple cubic is eight equivalent particles on the eight corners.

The body-centered cubic unit cell is the simplest repeating unit in a body-centered cubic structure. Once again, there are eight identical particles on the eight corners of the unit cell. However, this time there is a ninth identical particle in the center of the body of the unit cell.

The face-centered cubic unit cell also starts with identical particles on the eight corners of the cube. But this structure also contains the same particles in the centers of the six faces of the unit cell, for a total of 14 identical lattice points.

The face-centered cubic unit cell is the simplest repeating unit in a cubic closest-packed structure. In fact, the presence of face-centered cubic unit cells in this structure explains why the structure is known as cubic closest-packed.

Unit Cells: A Three-Dimensional Graph

The lattice points in a cubic unit cell can be described in terms of a three-dimensional graph. Because all three cell-edge lengths are the same in a cubic unit cell, it doesn't matter what orientation is used for a, b, and c axes. For the sake of argument, we'll define the axis as the vertical axis of our coordinate system, as shown in the figure below.

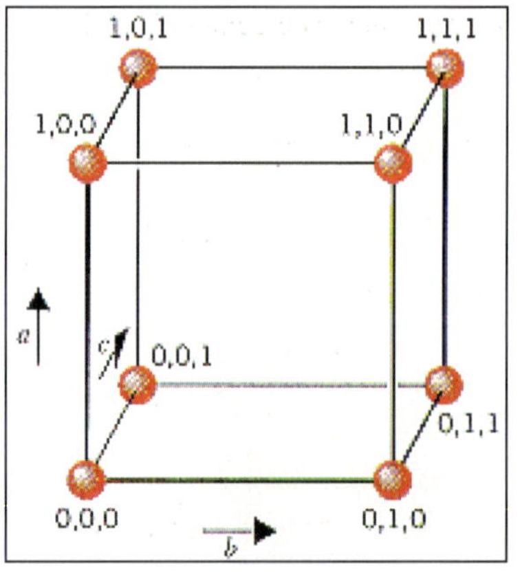

The b axis will then describe movement across the front of the unit cell, and the c axis will represent movement toward the back of the unit cell. Furthermore, we'll arbitrarily define the bottom left corner of the unit cell as the origin (0,0,0). The coordinates 1,0,0 indicate a lattice point that is one cell-edge length away from the origin along the a axis. Similarly, 0,1,0 and 0,0,1 represent lattice points that are displaced by one cell-edge length from the origin along the b and c axes, respectively.

Thinking about the unit cell as a three-dimensional graph allows us to describe the structure of a crystal with a remarkably small amount of information. We can specify the structure of cesium chloride, for example, with only four pieces of information.

- CsCl crystallizes in a cubic unit cell.
- The length of the unit cell edge is 0.4123 nm.
- There is a Cl^- ion at the coordinates 0,0,0.
- There is a Cs^+ ion at the coordinates 1/2,1/2,1/2.

Because the cell edge must connect equivalent lattice points, the presence of a Cl^- ion at one corner of the unit cell (0,0,0) implies the presence of a Cl^- ion at every corner of the cell. The coordinates 1/2,1/2,1/2 describe a lattice point at the center of the cell. Because there is no other point in the unit cell that is one cell-edge length away from these coordinates, this is the only Cs^+ ion in the cell. CsCl is therefore a simple cubic unit cell of Cl^- ions with a Cs^+ in the center of the body of the cell.

Unit Cells: NaCl and ZnS

NaCl should crystallize in a cubic closest-packed array of Cl^- ions with Na^+ ions in the octahedral holes between planes of Cl^- ions. We can translate this information into a unit-cell model for NaCl by remembering that the face-centered cubic unit cell is the simplest repeating unit in a cubic closest-packed structure.

There are four unique positions in a face-centered cubic unit cell. These positions are defined by the coordinates: 0,0,0; 0,1/2,1/2; 1/2,0,1/2; and 1/2,1/2,0. The presence of an particle at one corner of the unit cell (0,0,0) requires the presence of an equivalent particle on each of the eight corners of the unit cell. Because the unit-cell edge connects equivalent points, the presence of a particle in the center of the bottom face (0,1/2,1/2) implies the presence of an equivalent particle in the center of the top face (1,1/2,1/2). Similarly, the presence of particles in the center of the 1/2,0,1/2 and 1/2,1/2,0 faces of the unit cell implies equivalent particles in the centers of the 1/2,1,1/2 and 1/2,1/2,1 faces.

The figure below shows that there is an octahedral hole in the center of a face-centered cubic unit cell, at the coordinates 1/2,1/2,1/2. Any particle at this point touches the particles in the centers of the six faces of the unit cell.

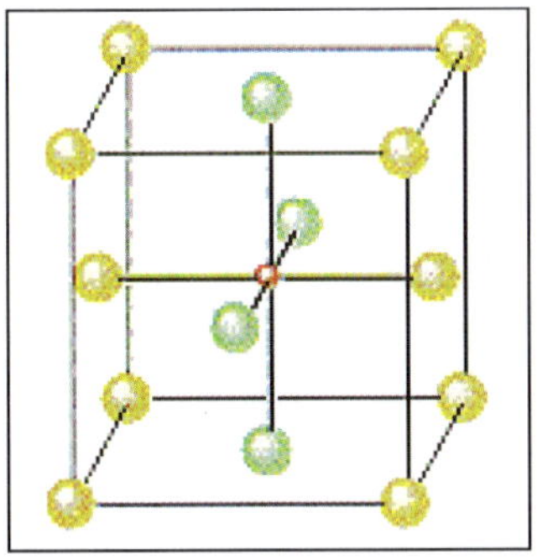

The other octahedral holes in a face-centered cubic unit cell are on the edges of the cell, as shown in the figure below.

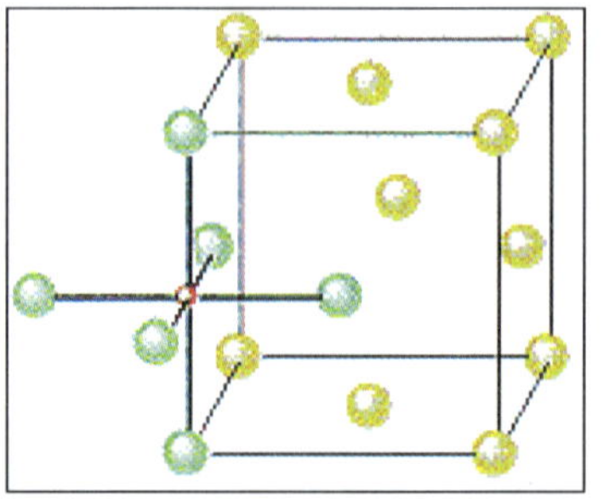

If Cl^- ions occupy the lattice points of a face-centered cubic unit cell and all of the octahedral holes are filled with Na^+ ions, we get the unit cell shown in the figure below.

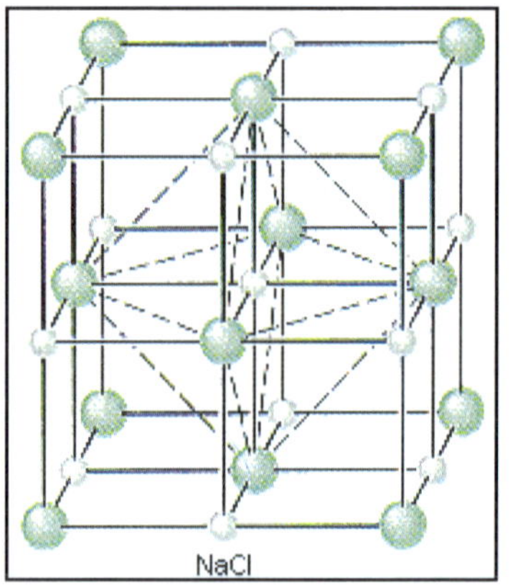

We can therefore describe the structure of NaCl in terms of the following information:

- NaCl crystallizes in a cubic unit cell.
- The cell-edge length is 0.5641 nm.
- There are Cl^- ions at the positions 0,0,0; 1/2,1/2,0; 1/2,0,1/2; and 0,1/2,1/2.
- There are Na^+ ions at the positions 1/2,1/2,1/2; 1/2,0,0; 0,1/2,0; and 0,0,1/2.

Placing a Cl^- ion at these four positions implies the presence of a Cl^- ion on each of the 14 lattice points that define a face-centered cubic unit. Placing a Na^+ ion in the center of the unit cell (1/2,1/2,1/2) and on the three unique edges of the unit cell (1/2,0,0; 0,1/2,0; and 0,0,1/2) requires an equivalent Na^+ ion in every octahedral hole in the unit cell.

ZnS crystallizes as cubic closest-packed array of S^{2-} ions with Zn^{2+} ions in tetrahedral holes. The S^{2-} ions in this crystal occupy the same positions as the Cl^- ions in NaCl. The only difference between these crystals is the location of the positive ions. The figure below shows that the tetrahedral holes in a face-centered cubic unit cell are in the corners of the unit cell, at coordinates such as 1/4,1/4,1/4. An atom with these coordinates would touch the atom at this corner as well as the atoms in the centers of the three faces that form this corner. Although it is difficult to see without a three-dimensional model, the four atoms that surround this hole are arranged toward the corners of a tetrahedron.

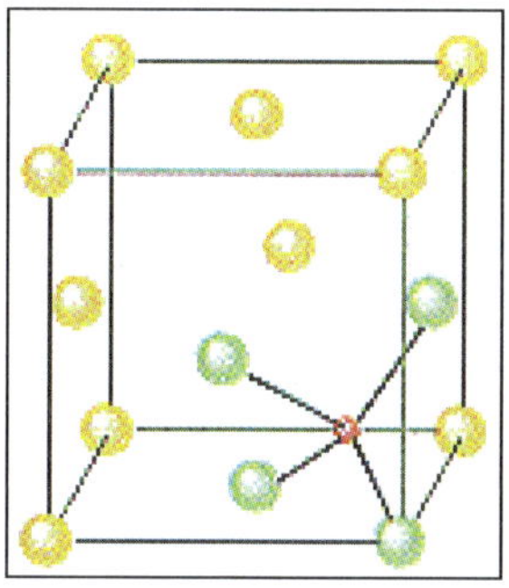

Because the corners of a cubic unit cell are identical, there must be a tetrahedral hole in each of the eight corners of the face-centered cubic unit cell. If S^{2-} ions occupy the lattice points of a face-centered cubic unit cell and Zn^{2+} ions are packed into every other tetrahedral hole, we get the unit cell of ZnS shown in the figure below.

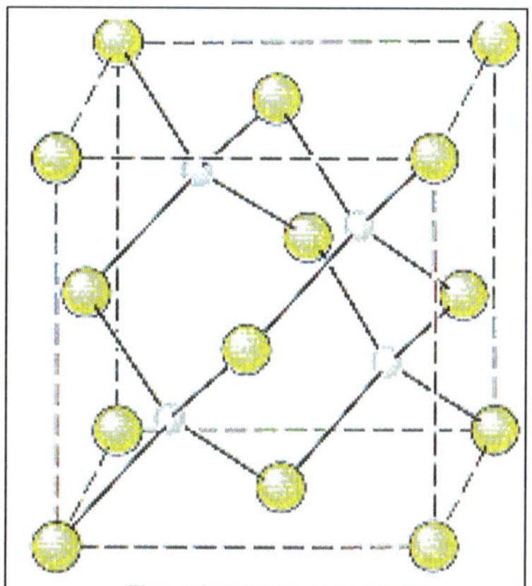
Zinc blende (cubic ZnS)

The structure of ZnS can therefore be described as follows:

- ZnS crystallizes in a cubic unit cell.
- The cell-edge length is 0.5411 nm.
- There are S^{2-} ions at the positions 0,0,0; 1/2,1/2,0; 1/2,0,1/2; and 0,1/2,1/2.
- There are Zn^{2+} ions at the positions 1/4,1/4,1/4; 1/4,3/4,3/4; 3/4,1/4,3/4; and 3/4,3/4,1/4.

The only half of the tetrahedral holes is occupied in this crystal because there are two tetrahedral holes for every S^{2-} ion in a closest-packed array of these ions.

Unit Cells: Measuring the Distance between Particles

Nickel is one of the metals that crystallize in a cubic closest-packed structure. When you consider that a nickel atom has a mass of only 9.75×10^{-23} g and an ionic radius of only 1.24×10^{-10} m, it is a remarkable achievement to be able to describe the structure of this metal. The obvious question is: How do we know that nickel packs in a cubic closest-packed structure?

The only way to determine the structure of matter on an atomic scale is to use a probe that is even smaller. One of the most useful probes for studying matter on this scale is electromagnetic radiation.

In 1912, Max van Laue found that x-rays that struck the surface of a crystal were diffracted into patterns that resembled the patterns produced when light passes through a very narrow slit. Shortly thereafter, William Lawrence Bragg, who was just completing his undergraduate degree in physics at Cambridge, explained van Laue's results with an equation known as the Bragg equation, which allows us to calculate the distance between planes of atoms in a crystal from the pattern of diffraction of x-rays of known wavelength:

$$n \lambda = 2d \sin T$$

The pattern by which x-rays are diffracted by nickel metal suggests that this metal packs in a cubic unit cell with a distance between planes of atoms of 0.3524 nm. Thus, the cell-edge length in this crystal must be 0.3524 nm. Knowing that nickel crystallizes in a cubic unit cell is not enough. We still have to decide whether it is a simple cubic, body-centered cubic or face-centered cubic unit cell. This can be done by measuring the density of the metal.

Unit Cells: Determining the Unit Cell of a Crystal

Atoms on the corners, edges, and faces of a unit cell are shared by more than one unit cell, as shown in the figure below. An atom on a face is shared by two unit cells, so only half of the atom belongs to each of these cells. An atom on an edge is shared by four unit cells, and an atom on a corner is shared by eight unit cells. Thus, only one-quarter of an atom on an edge and one-eighth of an atom on a corner can be assigned to each of the unit cells that share these atoms.

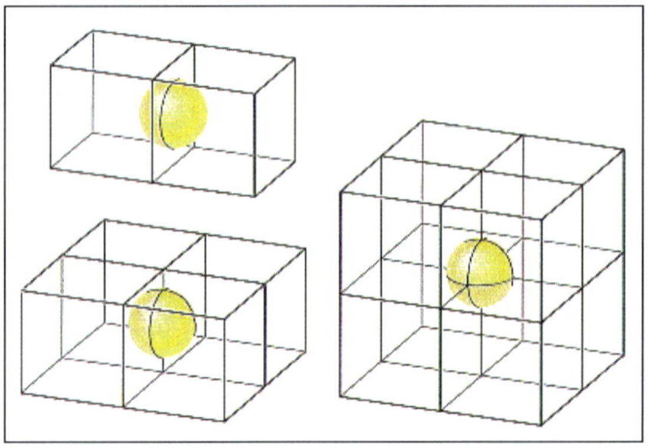

If nickel crystallized in a simple cubic unit cell, there would be a nickel atom on each of the eight corners of the cell. Because only one-eighth of these atoms can be assigned to a given unit cell, each unit cell in a simple cubic structure would have one net nickel atom:

Simple cubic structure → 8 corners x 1/8 = 1 atom.

If nickel formed a body-centered cubic structure, there would be two atoms per unit cell, because the nickel atom in the center of the body wouldn't be shared with any other unit cells:

Body-centered cubic structure → (8 corners x 1/8) + 1 body = 2 atoms.

If nickel crystallized in a face-centered cubic structure, the six atoms on the faces of the unit cell would contribute three net nickel atoms, for a total of four atoms per unit cell:

Face-centered cubic structure → (8 corners x 1/8) + (6 faces x 1/2) = 4 atoms.

Because they have different numbers of atoms in a unit cell, each of these structures would have a different density. Let's therefore calculate the density for nickel based on each of these structures and the unit cell edge length for nickel 0.3524 nm. In order to do this, we need to know the volume of the unit cell in cubic centimeters and the mass of a single nickel atom.

The volume (V) of the unit cell is equal to the cell-edge length (a) cubed:

$$V = a^3 = (0.3524\ nm)^3 = 0.04376\ nm^3$$

Since there are 10^9 nm in a meter and 100 cm in a meter, there must be 10^7 nm in a cm:

$$\frac{10^9}{1m} \times \frac{1m}{100\,cm} = 10^7\,nm/cm$$

We can therefore convert the volume of the unit cell to cm^3 as follows:

$$4.376 \times 10^{-2}\,nm^3 \times [\frac{1\,cm}{10^7\,nm}]^3 = 4.376 \times 10^{-23}\,cm^3$$

The mass of a nickel atom can be calculated from the atomic weight of this metal and Avogadro's number:

$$\frac{58.69\,g\,Ni}{1\,mol} \times \frac{1\,mol}{6.022 \times 10^{23}\,atoms} = 9.746 \times 10^{-23}\,glatom$$

The density of nickel, if it crystallized in a simple cubic structure, would therefore be 2.23 g/cm³, to three significant figures.

Simple cubic structure:

$$\frac{9.746 \times 10^{-23}\,g\,/\,unit\,cell}{4.376 \times 10^{-23}\,cm^3\,/\,unit\,cell} = 2.23\,g\,/\,cm^3$$

Because there would be twice as many atoms per unit cell if nickel crystallized in a body-centered cubic structure, the density of nickel in this structure would be twice as large.

Body-centered cubic structure:

$$\frac{2(9.746 \times 10^{-23}\,g\,/\,unit\,cell)}{4.376 \times 10^{-23}\,cm^3\,/\,unit\,cell} = 4.45\,g\,/\,cm^3$$

There would be four atoms per unit cell in a face-centered cubic structure and the density of nickel in this structure would be four times as large.

Face-centered cubic structure:

$$\frac{4(9.746 \times 10^{-23}\,g\,/\,unit\,cell}{4.376 \times 10^{-23}\,cm^3\,/\,unit\,cell} = 8.91\,g\,/\,cm^3$$

The experimental value for the density of nickel is 8.90 g/cm³. The obvious conclusion is that nickel crystallizes in a face-centered cubic unit cell and therefore has a cubic closest-packed structure.

Unit Cells: Calculating Metallic or Ionic Radii

Estimates of the radii of most metal atoms can be found. Where do these data come from? How do we know, for example, that the radius of a nickel atom is 0.1246 nm? Nickel crystallizes in a face-centered cubic unit cell with a cell-edge length of 0.3524 nm to calculate the radius of a nickel atom.

One of the faces of a face-centered cubic unit cell is shown in the figure.

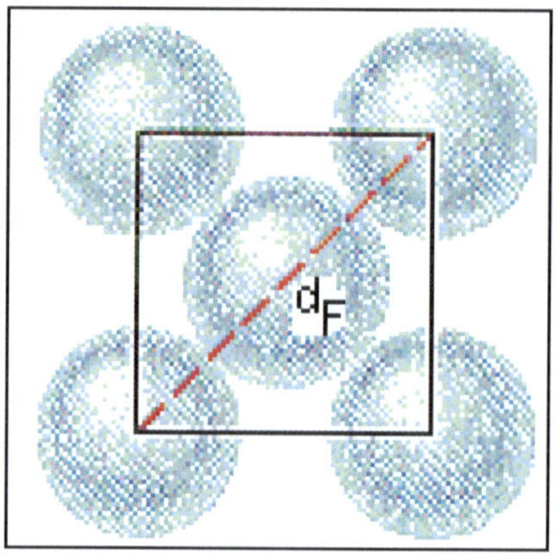

According to this figure, the diagonal across the face of this unit cell is equal to four times the radius of a nickel atom:

$$d_{face} = 4\gamma_N$$

The Pythagorean theorem states that the diagonal across a right triangle is equal to the sum of the squares of the other sides. The diagonal across the face of the unit cell is therefore related to the unit-cell edge length by the following equation:

$$d^2_{face} = a^2 + a^2$$

Taking the square root of both sides gives the following result:

$$d_{face} = \sqrt{2}\,a$$

We now substitute into this equation the relationship between the diagonal across the face of this unit cell and the radius of a nickel atom:

$$4\gamma_N = \sqrt{2}\,a$$

Solving for the radius of a nickel atom gives a value of 0.1246 nm:

$$\gamma_N = \frac{\sqrt{2}\,a}{4} = \frac{\sqrt{2}(0.3524\,\text{nm})}{4} = 0.1246\,nm$$

A similar approach can be taken to estimating the size of an ion. Let's start by using the fact that the cell-edge length in cesium chloride is 0.4123 nm to calculate the distance between the centers of the Cs^+ and Cl^- ions in CsCl.

CsCl crystallizes in a simple cubic unit cell of Cl^- ions with a Cs^+ ion in the center of the body of the cell, as shown in the figure below.

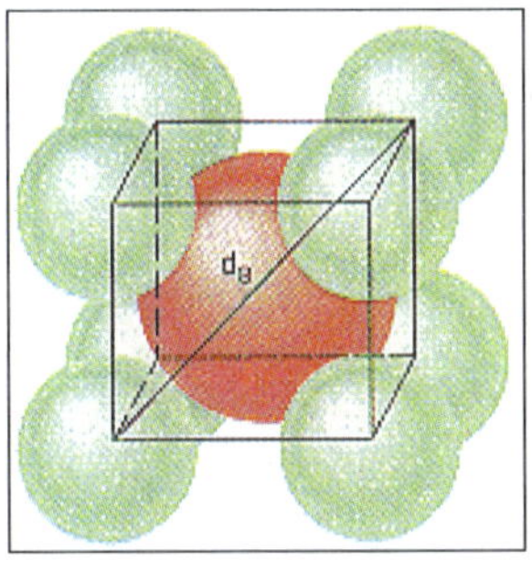

Before we can calculate the distance between the centers of the Cs^+ and Cl^- ions in this crystal, however, we have to recognize the validity of one of the simplest assumptions about ionic solids: The positive and negative ions that form these crystals touch.

We can therefore assume that the diagonal across the body of the CsCl unit cell is equivalent to the sum of the radii of two Cl^- ions and two Cs^+ ions:

$$d_{body} = 2\gamma_{cs+} + 2\gamma_{cl-}$$

The three-dimensional equivalent of the Pythagorean theorem suggests that the square of the diagonal across the body of a cube is the sum of the squares of the three sides:

$$d_{body}^2 = a^2 + a^2 + a^2$$

Taking the square root of both sides of this equation gives the following result:

$$d_{body} = \sqrt{3}\,a$$

If the cell-edge length in CsCl is 0.4123 nm, the diagonal across the body in this unit cell is 0.7141 nm:

$$d_{body} = \sqrt{3}\,a = \sqrt{3}(0.4123\,nm) = 0.7141\,nm$$

The sum of the ionic radii of Cs^+ and Cl^- ions is half this distance, or 0.3571 nm:

$$\gamma_{cs+} + \gamma_{cl-} = \frac{d_{body}}{2} = \frac{0.7141\,nm}{2} = 0.3571\,nm$$

If we had an estimate of the size of either the Cs^+ or Cl^- ion, we could use the results to calculate the radius of the other ion. The ionic radius of the Cl^- ion is 0.181 nm. Substituting this value into the last equation gives a value of 0.176 nm for the radius of the Cs^+ ion:

$$\gamma_{cs+} + \gamma_{cl-} = 0.3571\,nm$$
$$\gamma_{cs+} + 0.181\,nm = 0.3571\,nm$$

$$\gamma_{cs+} = 0.176\,nm$$

The results of this calculation are in reasonable agreement with the value of 0.169 nm known for the radius of the Cs^{+} ion. The discrepancy between these values reflects the fact that ionic radii vary from one crystal to another. The tabulated values are averages of the results of a number of calculations of this type.

Defects in a Solid

When we talk about crystalline solids, we say they have a definite structure. They have long term arrangement of their molecules. This means that they have repeated arrangement of a unit cell to form a space lattice structure. So the arrangement of molecules, in the long run, will remain the same.

However, these crystals are not perfect. They have defects or imperfections in their arrangement of constituent particles. These defects are a deviation from their uniform arrangement of particles.

The defects are of two kinds:

- Point Defects: When the deviation occurs around an atom/particle it is a point defect. It could be due to displacement, an extra particle or a missing particle.
- Line Defects: When there is an abnormality in the arrangement of an entire row, then it is a line defect.

Point Defects

Point defects usually occur when the crystallization has occurred very rapidly, not allowing a perfect crystal arrangement to form. Although it should be noted there are imperfections in even during a slow crystallization. There are mainly three different kinds of point defects.

Stoichiometric Defects

Stoichiometric compounds are those who maintain their stoichiometry. That means they maintain their ratios of cations and anions as indicated by their chemical formula. The defects in the solid do not affect this ratio.

Vacancy Defect

Here the lattice site is simply vacant, which means there is a missing particle. In a perfect crystal, there would not be this vacancy. This defect will lead to a reduced density of the solid. Some surrounding particle may move to fill in the gap, but the vacancy will only shift in the opposite direction, Also the solid structure of the crystal will ensure that the particles surrounding the vacant spot do not collapse.

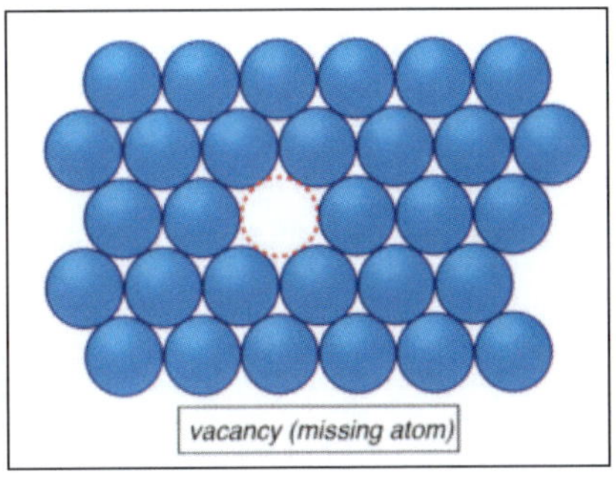

Interstitial Defect

There is an unoccupied space at the very center of the cube structure of the solid. When the eight spheres of a unit cell meet at the center they leave a little space, the interstitial site. Sometimes another particle will occupy this space. This is what we call an interstitial defect.

These are extra atoms or molecules or ions that occupy space which was supposed to be empty, which is why it is a defect. The density of the solid also increases due to such defects.

Schottky Defects

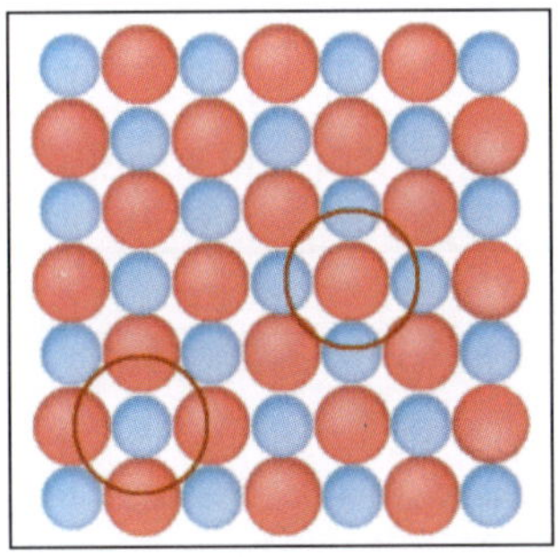

In this defect, more than one particle is missing. But the number of cations missing is equal to the number of anions missing. So the electrical composition remains the same.

Since more than one particle is missing, the mass of the structure will decrease. But since the volume remains the same, the density of the solid will also decrease. If there are too many particles missing, then the lattice structure may be compromised. This could mean the stability of the structure may suffer. An example of this defect is NaCl, KCl and other such ionic compounds.

Frenkel Defects

When a cation is missing from its normal position and has instead occupied an interstitial site it is a Frenkel Defect. Here the electrical neutrality will be unaffected. Also, it is usually cations that cause this defect since they can easily fit in the interstitial site due to their small and compact size.

Here no particles are actually missing, only the position is wrong. So the density of the solid will not change. It usually occurs in compounds where there is a significant size difference between cations and anions. Some examples are AgCl and AgBr etc.

Non-stoichiometric Defects

These defects will change the ration of ions in the solid. The number of cations and anions will not reflect as stated in their chemical formula. This imperfection is generally found in a large number of inorganic compounds. There are mainly two types of non-stoichiometry defects.

Metal Excess Defects

Here there is a surplus of metal ions in the space lattice. This can happen in one of two ways:

- Anionic Vacancy: Sometimes an anion may be missing from its usual position in the unit cell. This empty space will be occupied by an electron to maintain the electric neutrality of the solid. This space that the electron invades, is known as the F-center. This F-center electron is what gives the compound a color, one common example is NaCl and the yellow color it obtains on being heated.
- Extra Cations: On being heated many compounds result in having extra cations. These cations fit into the interstitial sites. To balance the neutrality an equal number of electrons do the same. But the ultimate result is the excess of metal in the solid. An example of this is when Zinc Oxide is heated and it losses oxygen.

Metal Deficiency Defects

In such compounds, there is less metal than its ideal stoichiometric proportions. Sometimes a positive ion is missing from the lattice site, but the neutrality of the solid will remain same as a neighbouring ion will take up two charges. The metal deficiency defect is normally found in Transition Elements that have the ability to have multiple valencies.

Atomic Structure

The atomic structure of an element refers to the constitution of its nucleus and the arrangement of the electrons around it. Primarily, the atomic structure of matter is made up of protons, electrons, and neutrons.

The protons and neutrons make up the nucleus of the atom, which is surrounded by the electrons belonging to the atom. The atomic number of an element describes the total number of protons in its nucleus.

Neutral atoms have equal numbers of protons and electrons. However, atoms may gain or lose electrons in order to increase their stability, and the resulting charged entity is called an ion.

Atoms of different elements have different atomic structures because they contain different numbers of protons and electrons. This is the reason for the unique characteristics of different elements.

Atomic Models

In the 18th and 19th centuries, many scientists attempted to explain the structure of the atom with the help of atomic models. Each of these models had their own merits and demerits and were pivotal to the development of the modern atomic model. The most notable contributions to the field were by the scientists John Dalton, J.J. Thomson, Ernest Rutherford, and Niels Bohr.

Dalton's Atomic Theory

The English chemist John Dalton suggested that all matter is made up of atoms, which were indivisible and indestructible. He also stated that all the atoms of an element were exactly the same, but the atoms of different elements differ in size and mass.

Chemical reactions, according to Dalton's atomic theory, involves a rearrangement of atoms to form products. According to the postulates proposed by Dalton, the atomic structure comprised of atoms, the smallest particle responsible for the chemical reactions to occur.

The following are the postulates of his theory:

- Every matter is made up of atoms.
- Atoms are indivisible.
- Specific elements have only one type of atoms in them.
- Each atom has its own constant mass that varies from element to element.
- Atoms undergo rearrangement during a chemical reaction.
- Atoms can neither be created nor be destroyed but can be transformed from one form to another.

Dalton's atomic theory successfully explained the Laws of chemical reactions, namely, the Law of conservation of mass, Law of constant properties, Law of multiple proportions, and Law of reciprocal proportions.

Demerits of Dalton's Atomic Theory

- The theory was unable to explain the existence of isotopes.
- Nothing about the structure of atom was appropriately is explained.
- Later, the scientists discovered particles inside the atom that proved, the atoms are divisible.

The discovery of particles inside atoms led to a better understanding of chemical species, these particles inside the atoms are called subatomic particles. The discovery of various subatomic particles is as follows:

Thomson Atomic Model

The English chemist Sir Joseph John Thomson put forth his model describing the atomic structure in the early 1900s. He was later awarded the Nobel prize for the discovery of "electrons". His work is based on an experiment called cathode ray experiment.

Cathode Ray Experiment

It has a tube made of glass which has two openings, one for the vacuum pump and the other for the inlet through which a gas is pumped.

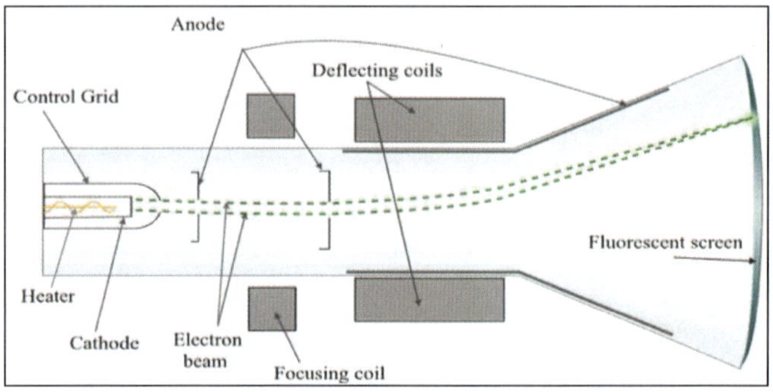

Cathode Ray Experiment – The Discovery of Electrons.

The role of the vacuum pump is to maintain "partial vacuum" inside the glass chamber. A high voltage power supply is connected using electrodes i.e. cathode and Anode is fitted inside the glass tube.

Observations

- When a high voltage power supply is switched on, there were rays emerging from the cathode towards the anode. This was confirmed by the 'Fluorescent spots' on the ZnS screen used. These rays were called "Cathode Rays".
- When an external electric field is applied, the cathode rays get deflected towards the positive electrode, but in the absence of electric field, they travel in a straight line.
- When rotor Blades are placed in the path of the cathode rays, they seem to rotate. This proves that the cathode rays are made up of particles of a certain mass so that they have some energy.

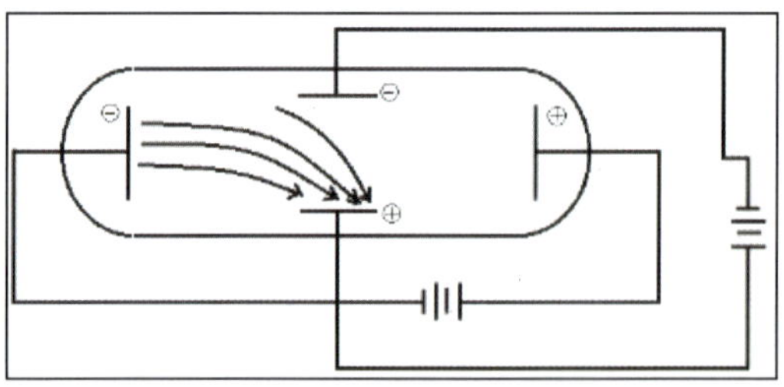

The direction of Cathode rays in the absence of an Electric Field.

- With all this evidence, Thompson concluded that cathode rays are made of negatively charged particle called "electrons".

On applying the electric and magnetic field upon the cathode rays (electrons), Thomson found the charge to mass ratio (e/m) of electrons. (e/m) for electron: 17588×10^{11} e/bg.

From this ratio, the charge of the electron was found by Mullikin through oil drop experiment. [Charge of e- = 1.6×10^{-16} C and Mass of e- = 9.1093×10^{-31} kg].

Result:

Based on results from his cathode ray experiment, Thomson described the atomic structure as a positively charged sphere into which negatively charged electrons were embedded.

It is commonly referred to as the "plum pudding model" because it can be visualized as a plum pudding dish where the pudding describes to the positively charged atom and the plum pieces describe the electrons.

Thomson's atomic structure described atoms as electrically neutral, i.e. the positive and the negative charges were of equal magnitude.

Limitations of Thomson's Atomic Structure: Thomson's atomic model does not clearly explain the stability of an atom. Also, further discoveries of other subatomic particles, couldn't be placed inside his atomic model.

Rutherford Atomic Theory

Rutherford, a student of J. J. Thomson modified the atomic structure with the discovery of another subatomic particle called "Nucleus". His atomic model is based on the Alpha ray scattering experiment.

Alpha Ray Scattering Experiment

Construction

- A very thin gold foil of 1000 atoms thick is taken.
- Alpha rays (doubly charged Helium He^{2+}) were made to bombard the gold foil.
- ZnS screen is placed behind the gold foil.

Observations

- Most of the rays just went through the gold foil making scintillations (bright spots) in the ZnS screen.
- A few rays got reflected after hitting the gold foil.
- One in 1000 rays got reflected by an angle of 180° (retraced path) after hitting the gold foil.

Result:

- Since most rays passed through, Rutherford concluded that most of the space inside the atom is empty.
- Few rays got reflected because of the repulsion of its positive with some other positive charge inside the atom.
- 1/1000th of rays got strongly deflected because of a very strong positive charge in the center of the atom he called this strong positive charge as "nucleus".
- He said most of the charge and mass of the atom resides in the Nucleus.

Rutherford's Structure of Atom

Based on the observations and results, Rutherford proposed his own atomic structure which is as follows:

- The nucleus is at the center of an atom, where most of the charge and mass are concentrated.
- Atomic structure is spherical.
- Electrons revolve around the nucleus in a circular orbit, similar to the way planets orbit the sun.

Limitations of Rutherford Atomic Model:

- If electrons have to revolve around the nucleus, they will spend energy and that too against the strong force of attraction from the nucleus, a lot of energy will be spent by the electrons and eventually, they will lose all their energy and will fall into the nucleus so the stability of atom is not explained.
- If electrons continuously revolve around the 'nucleus, the type of spectrum expected is a continuous spectrum. But in reality, what we see is a line spectrum.

Subatomic Particles

Protons

- Protons are positively charged subatomic particles. The charge of a proton is 1e, which corresponds to approximately 1.602×10^{-19}.
- The mass of a proton is approximately 1.672×10^{-24}.
- Protons are over 1800 times heavier than electrons.
- The total number of protons in the atoms of an element is always equal to the atomic number of the element.

Neutrons

- The mass of a neutron is almost the same as that of a proton i.e. 1.674×10^{-24}.
- Neutrons are electrically neutral particles and carry no charge.
- Different isotopes of an element have the same number of protons but vary in the number of neutrons present in their respective nuclei.

Electrons

- The charge of an electron is -1e, which approximates to -1.602×10^{-19}.
- The mass of an electron is approximately 9.1×10^{-31}.
- Due to the relatively negligible mass of electrons, they are ignored when calculating the mass of an atom.

Atomic Structure of Isotopes

Nucleons are the components of the nucleus of an atom. A nucleon can either be a proton or a neutron. Each element has a unique number of protons in it, which is described by its unique atomic number. However, several atomic structures of an element can exist, which differ in the total number of nucleons.

These variants of elements having a different nucleon number (also known as the mass number) are called isotopes of the element. Therefore, the isotopes of an element have the same number of protons but differ in the number of neutrons.

The atomic structure of an isotope is described with the help of the chemical symbol of the element, the atomic number of the element, and the mass number of the isotope. For example, there exist three known naturally occurring isotopes of hydrogen, namely, protium, deuterium, and tritium. The atomic structures of these hydrogen isotopes are illustrated below.

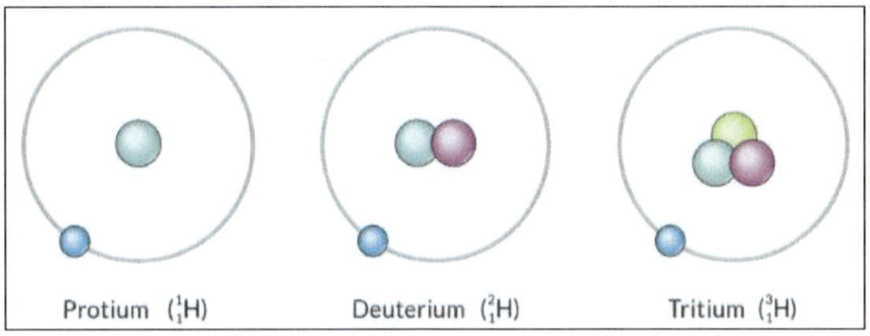

Atomic structures of hydrogen isotopes.

The isotopes of an element vary in stability. The half-lives of isotopes also differ. However, they generally have similar chemical behavior owing to the fact that they hold the same electronic structures.

Atomic Structures of Some Elements

The structure of atom of an element can be simply represented via the total number of protons, electrons, and neutrons present in it. The atomic structures of a few elements are illustrated below.

Hydrogen

The most abundant isotope of hydrogen on the planet Earth is protium. The atomic number and the mass number of this isotope are 1 and 1 respectively.

Structure of Hydrogen atom: This implies that it contains one proton, one electron, and no neutrons (total number of neutrons = mass number – atomic number).

Carbon

Carbon has two stable isotopes – 12C and 13C. Of these isotopes, 12C has an abundance of 98.9%. It contains 6 protons, 6 electrons, and 6 neutrons.

Structure of Carbon atom: The electrons are distributed into two shells and the outermost shell (valence shell) has four electrons. The tetravalency of carbon enables it to form a variety of chemical bonds with various elements.

Oxygen

There exist three stable isotopes of oxygen – 18O, 17O, and 16O. However, oxygen-16 is the most abundant isotope.

Structure of Oxygen atom: Since the atomic number of this isotope is 8 and the mass number is 16, it consists of 8 protons and 8 neutrons. 6 out of the 8 electrons in an oxygen atom lie in the valence shell.

Bohr's Atomic Theory

Neils Bohr put forth his model of the atom in the year 1915. This is the most widely used atomic model to describe the atomic structure of an element which is based on Planck's theory of quantization.

Postulates

- The electrons inside atoms are placed in discrete orbits called "stationery orbits".
- The energy levels of these shells can be represented via quantum numbers.
- Electrons can jump to higher levels by absorbing energy and move to lower energy levels by losing or emitting its energy.
- As long as, an electron stays in its own stationery, there will be no absorption or emission of energy.
- Electrons revolve around the nucleus in these stationery orbits only.
- The energy of the stationery orbits is quantized.

Limitations of Bohr's Atomic Theory:

- Bohr's atomic structure works only for single electron species such as H, He^+, Li_2^+, Be_3^+, etc.
- When the emission spectrum of hydrogen was observed under a more accurate spectrometer, each line spectrum was seen to be a combination of no of smaller discrete lines.
- Both Stark and Zeeman effects couldn't be explain using Bohr's theory.

Heisenberg's Uncertainty Principle

Heisenberg stated that no two conjugate physical quantities can be measured simultaneously with 100% accuracy. These will always be some error or uncertainty in the measurement.

- Drawback: Position and momentum are two such conjugate quantities which were measured accurately by Bohr (theoretically).
- Stark effect: Phenomenon of deflection of electrons in the presence of an electric field.
- Zeeman effect: Phenomenon of deflection of electrons in the presence of a magnetic field.

Dual Nature of Matter

The electrons which were treated to be particles, the evidence of photoelectric effect shows they also have wave nature. This was proved by Thomas young with the help of his double slit experiment.

De-Broglie concluded that since nature is symmetrical, so should be light or any other matter wave.

Quantum Numbers

- Principal Quantum number (n): It denotes the orbital number or shell number of electron.
- Azimuthal Quantum numbers (l): It denotes the orbital (sub-orbit) of the electron.
- Magnetic Quantum number: It denotes the number of energy states in each orbit.
- Spin Quantum number(s): It denotes the direction of spin, S = -½ = Anticlockwise and ½ = Clockwise.

Electronic Configuration of an Atom

The electrons have to be filled in the s, p, d, f in accordance with the following rule:

- Aufbau's principle: The filling of electrons should take place in accordance with the ascending order of energy of orbitals:
    - Lower energy orbital should be filled first and higher energy levels.
    - The energy of orbital α(p + l) value it two orbitals have same (n + l) value, E α n.
    - Ascending order of energy 1s, 2s, 2p, 3s, 3p, 4s, 3d, . . .
- Pauli's exclusion principle: No two electrons can have all the four quantum numbers to be the same or, if two electrons have to placed in an energy state they should be placed with opposite spies.
- Hund's rule of maximum multiplicity: In case of filling degenerate (same energy) orbitals, all the degenerate orbitals have to be singly filled first and then only pairing has to happen.

Molecular Structure and Bonding

VBT

Valence bond (VB) theory is a chemical bonding theory that explains the chemical bonding between two atoms. Like molecular orbital (MO) theory, it explains bonding using principles of quantum mechanics. According to valence bond theory, bonding is caused by the overlap of half-filled atomic orbitals. The two atoms share each other's unpaired electron to form a filled orbital to form a hybrid orbital and bond together. Sigma and pi bonds are part of valence bond theory.

Postulates of Valence Bond Theory

The important postulates of the valence bond theory are listed below:

- Covalent bonds are formed when two valence orbitals (half-filled) belonging to two different atoms overlap on each other. The electron density in the area between the two bonding atoms increases as a result of this overlapping, thereby increasing the stability of the resulting molecule.
- The presence of many unpaired electrons in the valence shell of an atom enables it to form multiple bonds with other atoms. The paired electrons present in the valence shell do not take participate in the formation of chemical bonds as per the valence bond theory.
- Covalent chemical bonds are directional and are also parallel to the region corresponding to the atomic orbitals that are overlapping.
- Sigma bonds and pi bonds differ in the pattern that the atomic orbitals overlap in, i.e. pi bonds are formed from sidewise overlapping whereas the overlapping along the axis containing the nuclei of the two atoms leads to the formation of sigma bonds.

The formation of sigma and pi bonds is illustrated below.

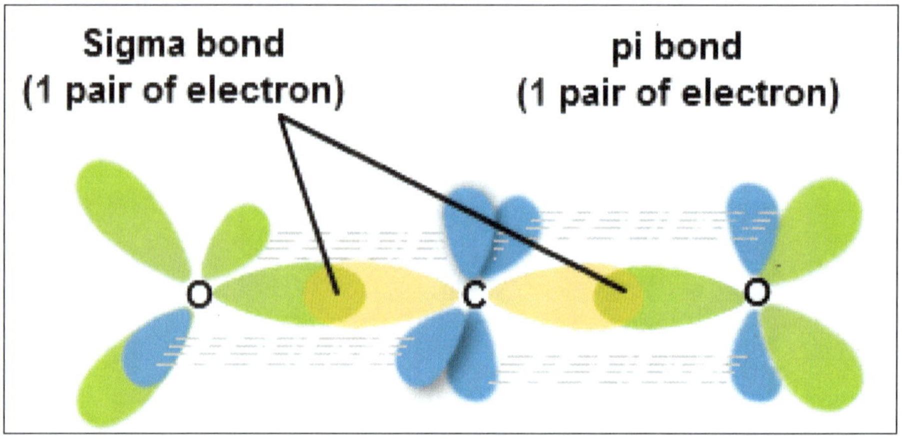

It can be noted that sigma bonds involve the head-to-head overlapping of atomic orbitals whereas pi bonds involve parallel overlapping.

Mechanism of Bonding in VB Theory

The VB theory describes the formation of covalent bonds from the overlap of atomic orbitals on two different atoms. Because of the overlap, it is highly probable that a pair of electrons are found in the physical region or space where the orbitals overlap.

Sigma (σ) and Pi (π) Bonds

There are two types of overlapping orbitals: sigma (σ) and pi (π). Both bonds are formed from the overlap of two orbitals, one on each atom. σ bonds occur when orbitals overlap between the nuclei of two atoms, also known as the internuclear axis.

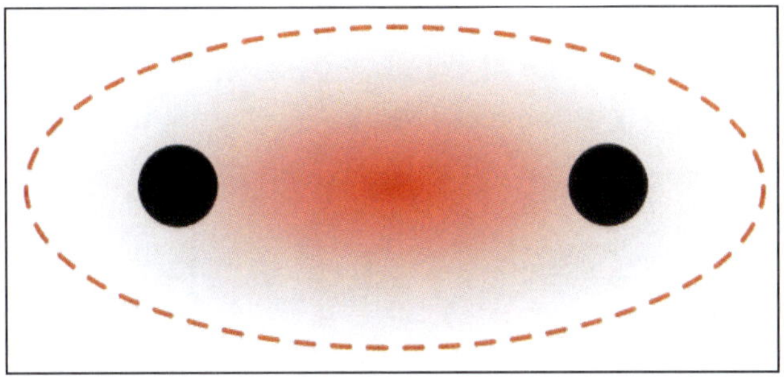

σ Bond Formation: Atomic orbitals from two atoms overlap in the region between the nuclei (internuclear axis). Therefore, the resulting electron density of the shared electrons lies in the red region shown in the image.

π bonds occur when two (unhybridized) p-orbitals overlap. The p-orbitals, in one π bond, are located above and below the nuclei of the atoms. By occupying the region of space that is above, below, and on the sides of an atom's nuclei, two π bonds can form.

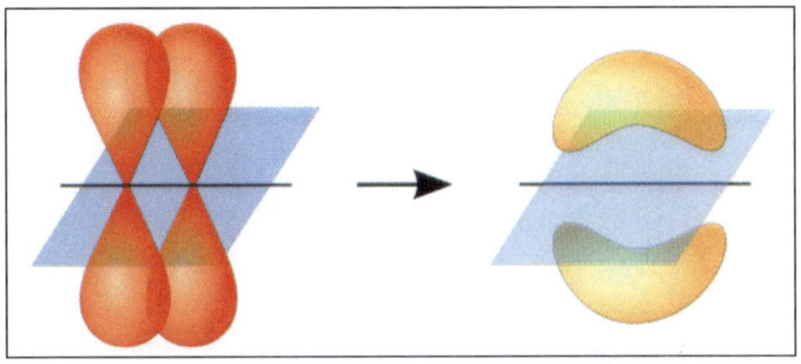

π Bond Formation: Two unhybridized p-orbitals can overlap so that the electron density of the shared electron pair is described by the π bond depicted on the right.

Both types of overlapping orbitals can be related to bond order. Single bonds have one sigma bond. Double bonds consist of one σ and one π bond, while triple bonds contain one σ and two π bonds.

Bond Character

An important aspect of the VB theory is the condition of maximum overlap which leads to the formation of the strongest possible bonds. This theory is used to explain the covalent bond formation in many molecules. In the F_2 molecule, the F–F σ covalent bond is formed by the overlap of p_z orbitals of the two F atoms, each containing an unpaired electron. Since the nature of the overlapping orbitals is different in H_2 and F_2 molecules, bond strength and bond lengths differ between H_2 and F_2 molecules.

In an HF molecule, the covalent σ bond forms from the overlap of the 1s orbital of H and the $2p_z$ orbital of F, each containing an unpaired electron. Mutual sharing of electrons between H and F results in a covalent bond in HF.

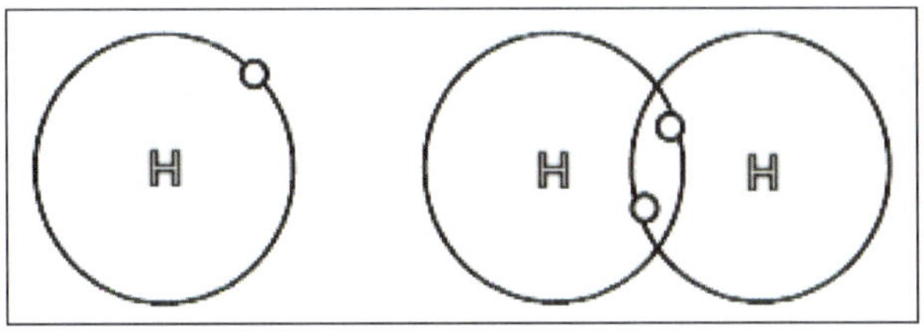

Covalent bond between hydrogen atoms: Each hydrogen atom has one electron. To complete their valence shells, they bond and share one electron with each other. This allows electrons to move about both atoms and gives both atoms access to two electrons; they become a stable H_2 molecule joined by a single covalent bond.

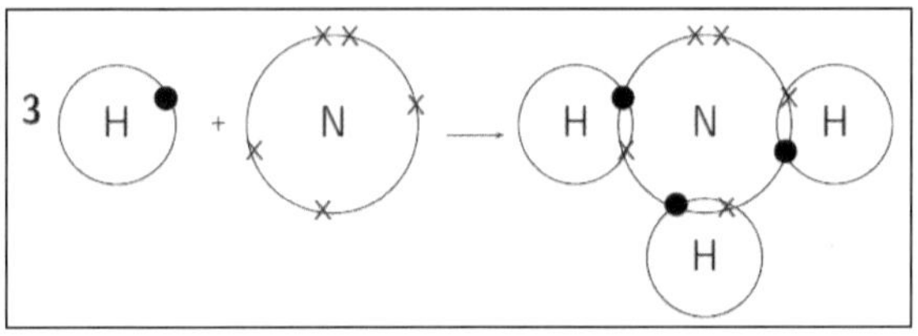

Covalent bonding in a molecule of ammonia: Each hydrogen atom needs one more electron to complete its valence energy shell. The nitrogen atom needs three more electrons to complete its valence energy shell. Therefore, three pairs of electrons must be shared between the four atoms involved. The nitrogen atom will share three of its electrons so that each of the hydrogen atoms now has a complete valence shell. Each of the hydrogen atoms will share its electron with the nitrogen atom to complete its valence shell.

sp^3 Hybridization

sp^3 hybrid orbitals form when a single s and three p orbitals hybridize.

In a tetravalent molecule, four outer atoms are bonded to a central atom. Perhaps the most common and important example of this bond type is methane, CH_4. In the ground state of the free carbon atom, there are two unpaired electrons in separate 2p orbitals. To form four bonds, the atom must have four unpaired electrons; this requires that carbon's valence 2s and 2p orbitals each contain an electron for bonding.

The single 2s orbital is spherical, different from the dumbbell-shaped 2p orbitals. This would indicate that one of the four bonds differs from the other three, but scientific tests have proven that all four bonds have equal length and energy; this is due to the hybridization of carbon's 2s and 2p valence orbitals.

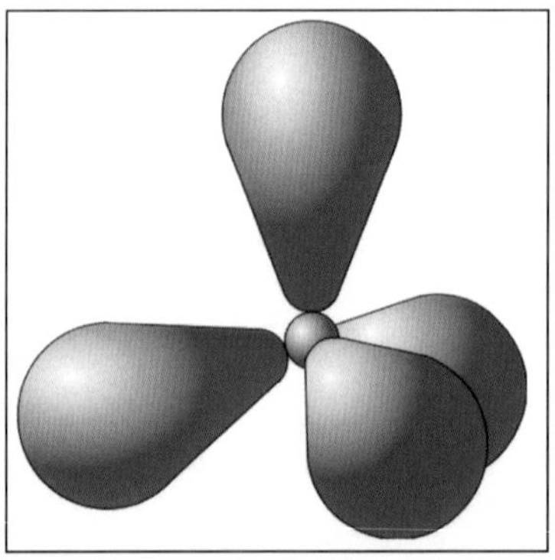

Methane: The methane molecule has four equal bonds.

In hybridization, carbon's 2s and three 2p orbitals combine into four identical orbitals, now called sp^3 hybrids.

The bonds between carbon and hydrogen can form the backbone of very complicated and extensive chain hydrocarbon molecules. The simplest of these is ethane (C_2H_6), in which an sp^3 orbital on each of the two carbon atoms joins (overlaps) to form a carbon-carbon bond; then, the remaining carbon sp^3 orbital overlaps with six hydrogen 1s orbitals to form the ethane molecule.

Ethane: Ethane can form by replacing one of the hydrogen atoms in CH_4 with another sp_3 hybridized carbon fragment.

If lone electron pairs are present on the central atom, that can occupy one or more of the sp^3 orbitals. For example, in the ammonia molecule, the fourth of the sp^3 hybrid orbitals on the nitrogen contains the two remaining outer-shell electrons, which form a non-bonding lone pair.

In the water molecule, the oxygen atom can form four sp^3 orbitals. Two of these are occupied by the two lone pairs on the oxygen atom, while the other two are used for bonding. The observed H-O-H bond angle in water (104.5°) is less than the tetrahedral angle (109.5°); one explanation for this is that the non-bonding electrons tend to remain closer to the central atom and thus exert greater repulsion on the other orbitals, pushing the two bonding orbitals closer together.

sp^2 Hybridization

sp^2 hybridization occurs between one s-orbital and two p-orbitals.

Boron trifluoride (BF_3) has a boron atom with three outer-shell electrons in its normal or ground state, as well as three fluorine atoms, each with seven outer electrons. One of the three boron electrons is unpaired in the ground state. In order to explain the bonding, the 2s orbital and two of the 2p orbitals (called sp^2 hybrids) hybridize; one empty p-orbital remains.

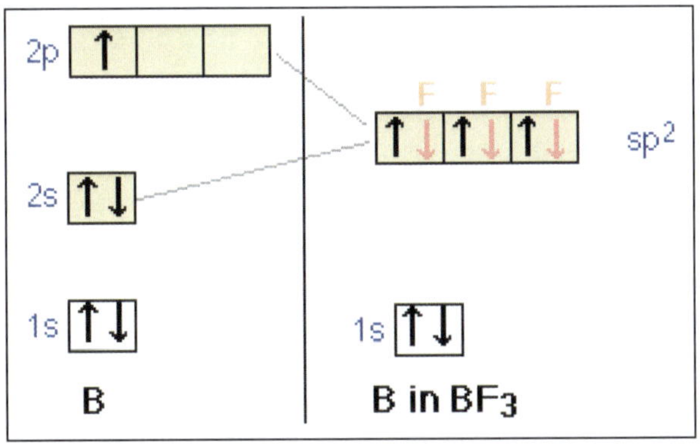

Boron configuration diagram: One of the three boron electrons is unpaired in its ground state. The atomic s- and p-orbitals in boron's outer shell mix to form three equivalent hybrid orbitals. These particular orbitals are called sp² hybrids, meaning that this set of orbitals derives from one s- orbital and two p-orbitals of the free atom.

sp² Hybridization in Ethene and the Formation of a Double Bond

Ethene (C_2H_4) has a double bond between the carbons. In this case, carbon will sp² hybridize; in sp² hybridization, the 2s orbital mixes with only two of the three available 2p orbitals, forming a total of three sp hybrid orbitals with one p-orbital remaining. The three hybridized orbitals explain the three sigma bonds that each carbon forms.

C^*	↑↓	↑	↑	↑	↑
	$1s$	sp^2	sp^2	sp^2	p

sp² hybridization in ethene: In sp² hybridization, the 2s orbital mixes with only two of the three available 2p orbitals, forming a total of three sp² orbitals with one p-orbital remaining.

The two carbon atoms form a sigma bond in the molecule by overlapping two sp² orbitals. Each carbon atom forms two covalent bonds with hydrogen by s–sp² overlap, all with 120° angles. The pi bond between the carbon atoms perpendicular to the molecular plane is formed by 2p–2p overlap.

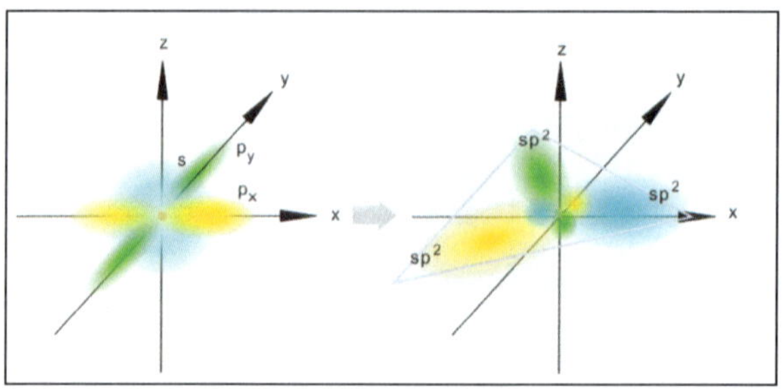

Formation of sp² hybrid orbitals: This illustration shows how an s-orbital mixes with two p orbitals to form a set of three sp² hybrid orbitals.

sp Hybridization

sp hybrid orbitals form from one s-orbital and one p-orbital. In sp hybridization, the s orbital overlaps with only one p orbital. Atoms that exhibit sp hybridization have sp orbitals that are linearly oriented; two sp orbitals will be at 180 degrees to each other.

Any central atom surrounded by just two regions of valence electron density in a molecule will exhibit sp hybridization. Some examples include the mercury atom in the linear $HgCl_2$ molecule, the zinc atom in $Zn(CH_3)_2$, which contains a linear C–Zn–C arrangement, the carbon atoms in HCCH and CO_2, and the Be atom in $BeCl_2$.

sp Hybridization in $BeCl_2$

The beryllium atom in a gaseous $BeCl_2$ molecule is an example of a central atom with no lone pairs of electrons in a linear arrangement of three atoms. There are two regions of valence electron density in the $BeCl_2$ molecule that correspond to the two covalent Be–Cl bonds. To accommodate these two electron domains, two of the Be atom's four valence orbitals will mix to yield two hybrid orbitals.

When atomic orbitals hybridize, the valence electrons occupy the newly created orbitals. The Be atom had two valence electrons, so each of the sp orbitals gets one of these electrons. Each of these electrons pairs up with the unpaired electron on a chlorine atom when a hybrid orbital and a chlorine orbital overlap during the formation of the Be–Cl bonds.

The hybridization process involves mixing of the valence s orbital with one of the valence p orbitals to yield two equivalent sp hybrid orbitals that are oriented in a linear geometry.

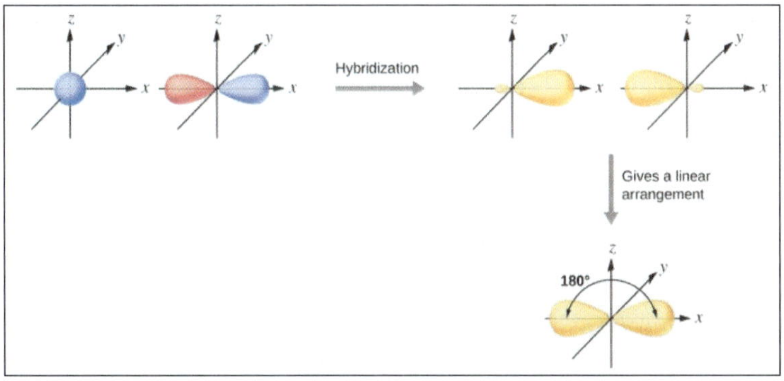

s and p orbitals combining to form two sp orbitals.

Hybridization of an s orbital and a p orbital of the same atom produces two sp hybrid orbitals. Each hybrid orbital is oriented primarily in just one direction. Each sp orbital contains one lobe that is significantly larger than the other. The set of two sp orbitals are oriented at 180°, which is consistent with the geometry for two domains.

The set of sp orbitals appears similar in shape to the original p orbital, but there is an important difference. The number of atomic orbitals combined always equals the number of hybrid orbitals formed. The p orbital is one orbital that can hold up to two electrons. The sp set is two equivalent orbitals that point 180° from each other. The two electrons that were originally in the s orbital are now distributed to the two sp orbitals, which are half filled.

In gaseous $BeCl_2$, these half-filled hybrid orbitals will overlap with orbitals from the chlorine atoms to form two identical σ bonds.

Energy Level Diagrams for sp Orbitals

The electronic differences in an isolated Be atom and in the bonded Be atom can be illustrated using an orbital energy-level diagram. These diagrams represent each orbital by a horizontal line (indicating its energy) and each electron by an arrow. Energy increases toward the top of the diagram. We use one upward arrow to indicate one electron in an orbital and two arrows (up and down) to indicate two electrons of opposite spin.

The left orbital energy-level diagram shows both electrons of Be in the 2s orbital. The right energy-level diagrams shows sp hybridized orbitals on Be in the linear $BeCl_2$ molecule. Each of the two sp hybrid orbitals holds one electron and is thus half filled and available for bonding via overlap with a Cl 3p orbital.

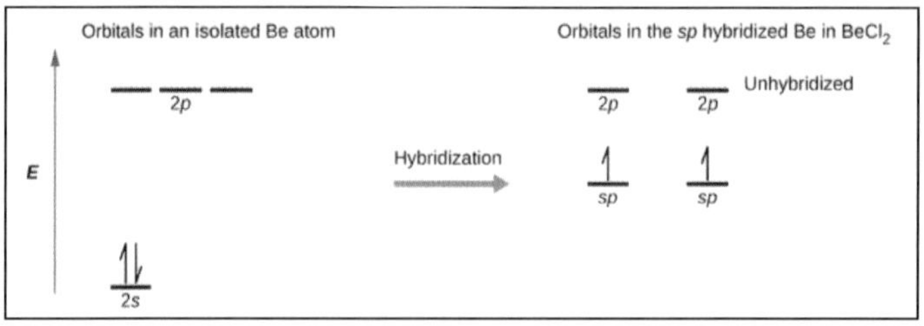

Orbital energy-level diagrams for isolated and bonded Be.

Hybridization in Molecules Containing Double and Triple Bonds

sp^2, sp hybridizations, and pi-bonding can be used to describe the chemical bonding in molecules with double and triple bonds.

In chemistry, hybridization is the concept of mixing atomic orbitals to form new hybrid orbitals suitable for describing bonding properties. Hybridized orbitals are very useful in explaining of the shape of molecular orbitals for molecules, and are an integral part of valence bond theory.

The hybrids are named for the atomic orbitals involved in the hybridization. In methane (CH_4) for example, a set of sp^3 orbitals forms by mixing one s- and three p-orbitals on the carbon atom. The orbitals are directed toward the four hydrogen atoms, which are located at the vertices of a regular tetrahedron.

Ethene (C_2H_4) has a double bond between the carbons. For this molecule, carbon will sp^2 hybridize. In sp^2 hybridization, the 2s orbital mixes with only two of the three available 2p orbitals, forming a total of 3 sp^2 orbitals with one p-orbital remaining. In ethylene (ethene), the two carbon atoms form a sigma bond by overlapping two sp^2 orbitals; each carbon atom forms two covalent bonds with hydrogen by s–sp^2 overlapping all with 120° angles. The pi bond between the carbon atoms forms by a 2p-2p overlap. The hydrogen-carbon bonds are all of equal strength and length, which agrees with experimental data.

Multiple bonds can also occur between dissimilar atoms. When the two O-atoms are brought up to opposite sides of the carbon atom in carbon dioxide, one of the p orbitals on each oxygen forms a pi bond with one of the carbon p-orbitals. In this case, sp hybridization leads to two double bonds.

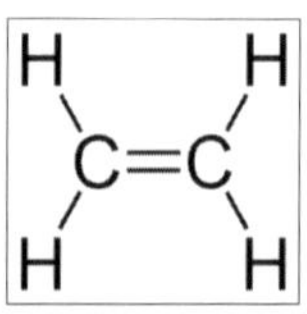

- Ethene structure: Ethene has a double bond between the carbons.

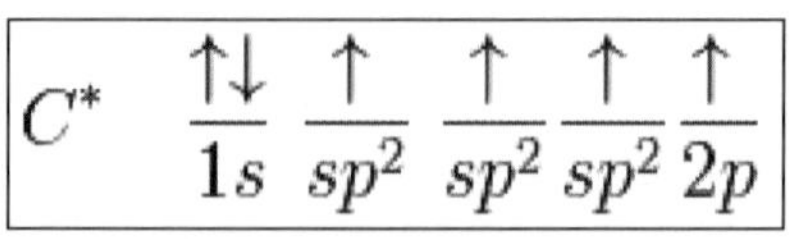

- sp^2 hybridization: In ethene, carbon sp^2 hybridizes, because one π (pi) bond is required for the double bond between the carbons, and only three σ bonds form per carbon atom.

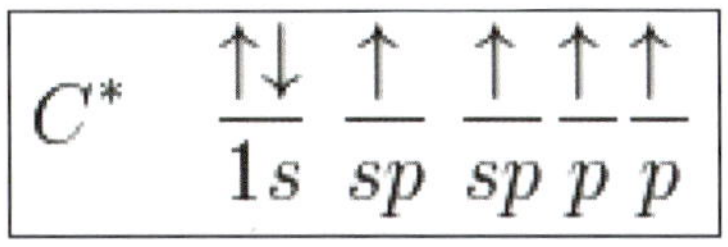

- sp hybridisation: In this model, the 2s orbital mixes with only one of the three p-orbitals, resulting in two sp-orbitals and two remaining unchanged p-orbitals.

sp hybridization explains the chemical bonding in compounds with triple bonds, such as alkynes; in this model, the 2s orbital mixes with only one of the three p-orbitals, resulting in two sp orbitals and two remaining p-orbitals. The chemical bonding in acetylene (ethyne) (C_2H_2) consists of sp-sp overlap between the two carbon atoms forming a sigma bond, as well as two additional pi bonds formed by p-p overlap. Each carbon also bonds to hydrogen in a sigma s-sp overlap at 180° angles.

Applications of VBT

The maximum overlap condition which is described by the valence bond theory can explain the formation of covalent bonds in several molecules. This is one of its most important applications. For example, the difference in the length and strength of the chemical bonds in H_2 and F_2 molecules can be explained by the difference in the overlapping orbitals in these molecules.

The covalent bond in an HF molecule is formed from the overlap of the 1s orbital of the hydrogen atom and a 2p orbital belonging to the fluorine atom, which is explained by the valence bond theory.

Limitations of Valence Bond Theory

The shortcomings of the valence bond theory include:

- Failure to explain the tetravalency exhibited by carbon.
- No insight offered on the energies of the electrons.
- The theory assumes that electrons are localized in specific areas.

MOT

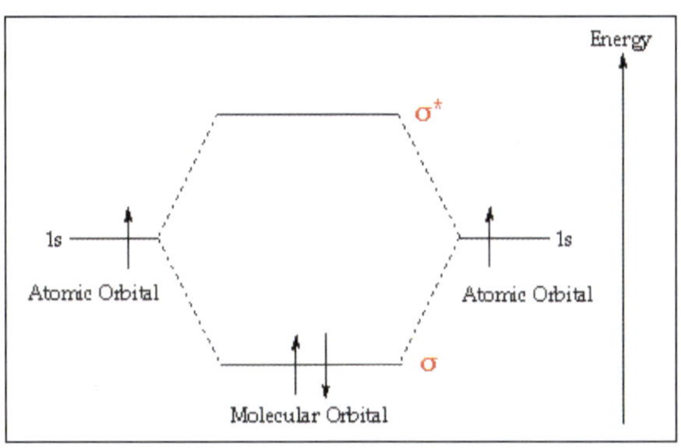

The Valence Bond Theory fails to answer certain questions like why He_2 molecule does not exist and why O_2 is paramagnetic.

According to the Molecular Orbital Theory, individual atoms combine to form molecular orbitals. Thus the electrons of an atom are present in various atomic orbitals and are associated with several nuclei.

We know that we can consider electrons as either particle or wave nature. Therefore, we can describe an electron in an atom as occupying an atomic orbital, or by a wave function Ψ. These are solutions to the Schrodinger wave equation. Electrons in a molecule occupy molecular orbitals. We can obtain the wave function of a molecular orbital by the following methods:

- Linear Combination of Atomic Orbitals (LCAO).
- United Atom Method.

Linear Combination of Atomic Orbitals (LCAO)

As per this method, the formation of orbitals is because of Linear Combination (addition or subtraction) of atomic orbitals which combine to form the molecule. Consider two atoms A and B which have atomic orbitals described by the wave functions Ψ_A and Ψ_B.

If the electron cloud of these two atoms overlaps, then we can obtain the wave function for the molecule by a linear combination of the atomic orbitals Ψ_A and Ψ_B. The below equation forms two molecular orbitals.

$$\Psi_{MO} = \Psi_A \pm \Psi_B$$

Bonding Molecular Orbitals

When the addition of wave function takes place, the type of molecular orbitals formed are Bonding Molecular Orbitals. We can represent them by $\Psi_{MO} = \Psi_A + \Psi_B$. They have lower energy than atomic orbitals involved.

Anti-Bonding Molecular Orbitals

When molecular orbital forms by the subtraction of wave function, the type of molecular orbitals formed are antibonding Molecular Orbitals. We can represent them as $\Psi_{MO} = \Psi_A - \Psi_B$. They have higher energy than atomic orbitals. Therefore, the combination of two atomic orbitals results in the formation of two molecular orbitals. They are the bonding molecular orbital (BMO) and the anti-bonding molecular orbital (ABMO).

Relative energies of molecular orbitals:

- Bonding Molecular Orbitals (BMO) – Energy of Bonding Molecular Orbitals is less than that of Anti Bonding Molecular Orbitals. This is because of the increase in the attraction of both the nuclei for both the electron (of the combining atom).
- Anti-Bonding Molecular Orbitals (ABMO) – Energy of Anti Bonding Molecular Orbitals is higher than Bonding Molecular Orbitals. This is because the electron tries to move away from the nuclei and are in a repulsive state.

The molecular orbital model is by far the most productive of the various models of chemical bonding, and serves as the basis for most quantiative calculations, including those that lead to many of the computer-generated images. In its full development, molecular orbital theory involves a lot of complicated mathematics, but the fundamental ideas behind it are quite easily understood.

This is a big departure from the simple Lewis and VSEPR models that were based on the one-center orbitals of individual atoms. The more sophisticated hybridization model recognized that these orbitals will be modified by their interaction with other atoms. But all of these valence-bond models, as they are generally called, are very limited in their applicability and predictive power, because they fail to recognize that distribution of the pooled valence electrons is governed by the totality of positive centers.

Molecular Orbitals

Chemical bonding occurs when the net attractive forces between an electron and two nuclei exceeds the electrostatic repulsion between the two nuclei. For this to happen, the electron must be in a region of space which we call the binding region. Conversely, if the electron is off to one side, in an anti-binding region, it actually adds to the repulsion between the two nuclei and helps push them away.

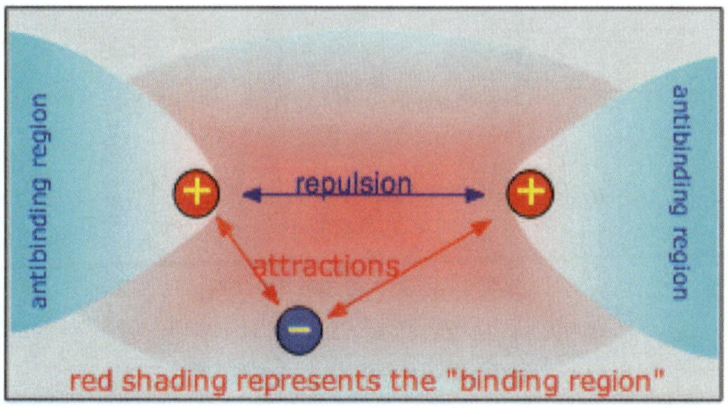

The easiest way of visualizing a molecular orbital is to start by picturing two isolated atoms and the electron orbitals that each would have separately. These are just the orbitals of the separate atoms, by themselves, which we already understand. We will then try to predict the manner in which these atomic orbitals interact as we gradually move the two atoms closer together. Finally, we will reach some point where the internuclear distance corresponds to that of the molecule. The corresponding orbitals will then be the molecular orbitals of our new molecule.

Bonding and Antibonding Molecular Orbitals

There is one minor difficulty we started with two orbitals (the 1s atomic orbitals), and ended up with only one orbital. Now according to the rules of quantum mechanics, orbitals cannot simply appear and disappear at our convenience. For one thing, this would raise the question of at just what internuclear distance do we suddenly change from having two orbitals, to having only one? It turns out that when orbitals interact, they are free to change their forms, but there must always be the same number. This is just another way of saying that there must always be the same number of possible allowed sets of electron quantum numbers.

How can we find the missing orbital? We must go back to the wave-like character of orbitals that we developed in our earlier treatment of the hydrogen atom. You are probably aware that wave phenomena such as sound waves, light waves, or even ocean waves can combine or interact with one another in two ways: they can either reinforce each other, resulting in a stronger wave, or they can interfere with and partially destroy each other. A roughly similar thing occurs when the "matter waves" corresponding to the two separate hydrogen 1s orbitals interact; both in-phase and out-of-phase combinations are possible, and both occur. The in-phase, reinforcing interaction yields the bonding orbital that we just considered. The other, corresponding to out-of-phase combination of the two orbitals, gives rise to a molecular orbital that has its greatest electron probability in what is clearly the antibonding region of space. This second orbital is therefore called an antibonding orbital.

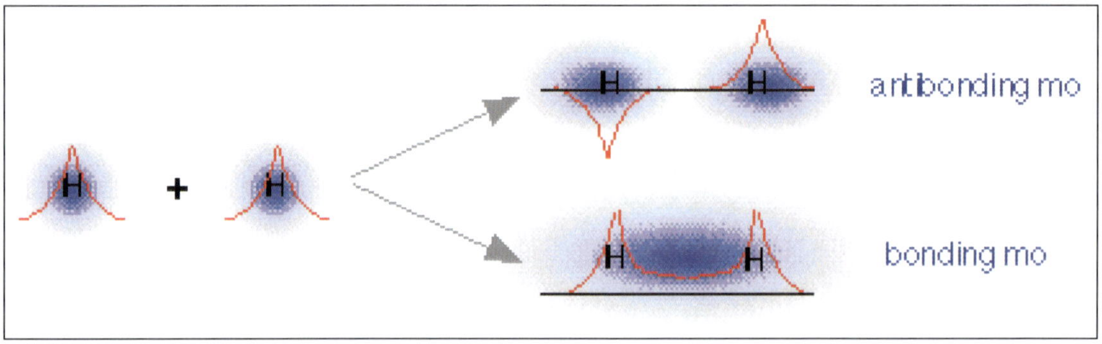

When the two 1s wave functions combine out-of-phase, the regions of high electron probability do not merge. In fact, the orbitals act as if they actually repel each other. Notice particularly that there is a region of space exactly equidistant between the nuclei at which the probability of finding the electron is zero. This region is called a nodal surface, and is characteristic of antibonding orbitals. It should be clear that any electrons that find themselves in an antibonding orbital cannot possibly contribute to bond formation; in fact, they will actively oppose it.

Two orbitals are originally on separate atoms, begin to interact as we push the two nuclei toward each other, these two atomic orbitals will gradually merge into a pair of molecular orbitals, one of which will have bonding character, while the other will be antibonding. In a more advanced treatment, it would be fairly easy to show that this result follows quite naturally from the wave-like nature of the combining orbitals.

What is the difference between these two kinds of orbitals, as far as their potential energies are concerned? More precisely, which kind of orbital would enable an electron to be at a lower potential energy? Clearly, the potential energy decreases as the electron moves into a region that enables it to see the maximum amount of positive charge. In a simple diatomic molecule, this will be in the internuclear region— where the electron can be simultaneously close to two nuclei. The bonding orbital will therefore have the lower potential energy.

Molecular Orbital Diagrams

This scheme of bonding and antibonding orbitals is usually depicted by a molecular orbital diagram such as the one shown here for the dihydrogen ion H_2^+. Atomic valence electrons (shown in boxes on the left and right) fill the lower-energy molecular orbitals before the higher ones, just as is the case for atomic orbitals. Thus, the single electron in this simplest of all molecules goes into the bonding orbital, leaving the antibonding orbital empty.

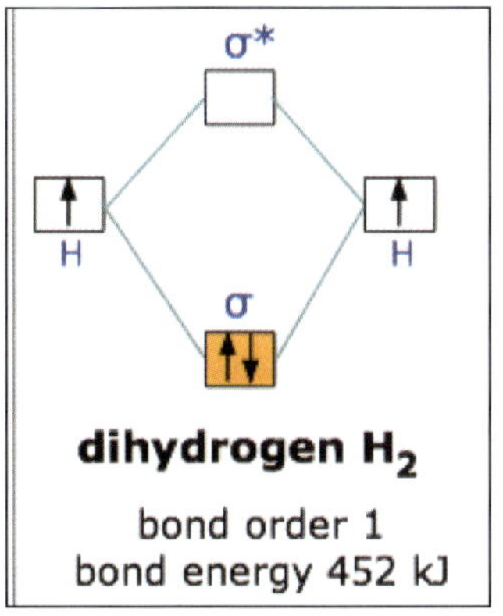

Since any orbital can hold a maximum of two electrons, the bonding orbital in H_2^+ is only half-full. This single electron is nevertheless enough to lower the potential energy of one mole of hydrogen nuclei pairs by 270 kJ— quite enough to make them stick together and behave like a distinct molecular species. Although H_2^+ is stable in this energetic sense, it happens to be an extremely reactive molecule— so much so that it even reacts with itself, so these ions are not commonly encountered in everyday chemistry.

Dihydrogen

If one electron in the bonding orbital is conducive to bond formation, might two electrons be even better? We can arrange this by combining two hydrogen atoms- two nuclei, and two electrons. Both electrons will enter the bonding orbital.

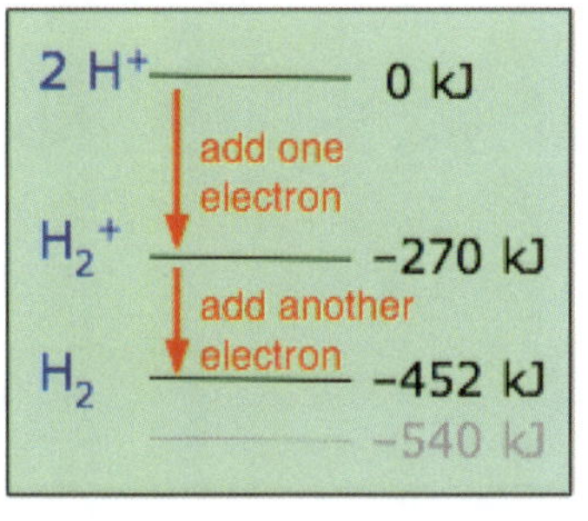

We recall that one electron lowered the potential energy of the two nuclei by 270 kJ/mole, so we might expect two electrons to produce twice this much stabilization, or 540 kJ/mole.

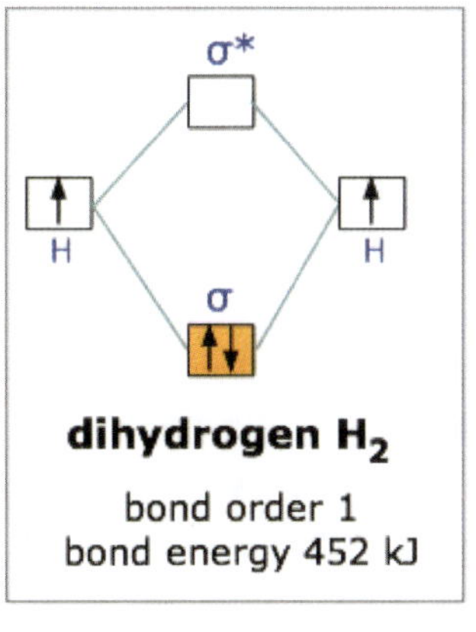

Bond order is defined as the difference between the number of electron pairs occupying bonding and nonbonding orbitals in the molecule. A bond order of unity corresponds to a conventional "single bond".

Experimentally, one finds that it takes only 452 kJ to break apart a mole of hydrogen molecules. The reason the potential energy was not lowered by the full amount is that the presence of two electrons in the same orbital gives rise to a repulsion that acts against the stabilization. This is exactly the same effect we saw in comparing the ionization energies of the hydrogen and helium atoms.

Dihelium

With two electrons we are still ahead, so let's try for three. The dihelium positive ion is a three-electron molecule. We can think of it as containing two helium nuclei and three electrons. This molecule is stable, but not as stable as dihydrogen; the energy required to break He_2^+ is 301 kJ/mole. The reason for this should be obvious; two electrons were accommodated in the bonding orbital, but the third electron must go into the next higher slot— which turns out to be the sigma antibonding orbital. The presence of an electron in this orbital, as we have seen, gives rise to a repulsive component which acts against, and partially cancels out, the attractive effect of the filled bonding orbital.

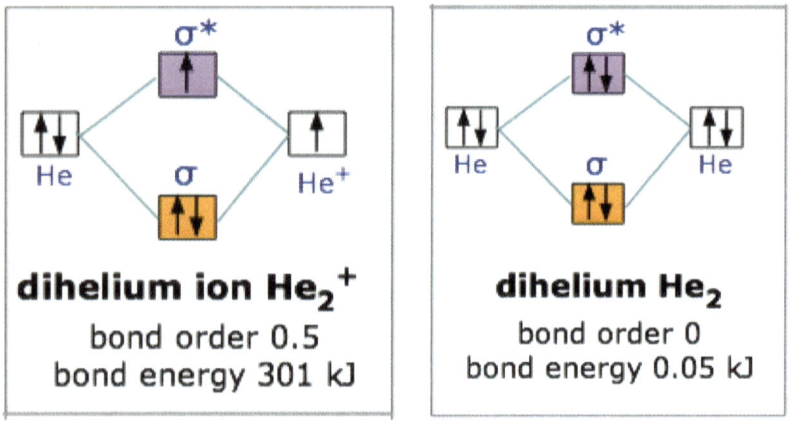

Taking our building-up process one step further, we can look at the possibilities of combining to helium atoms to form dihelium. You should now be able to predict that He_2 cannot be a stable molecule; the reason, of course, is that we now have four electrons— two in the bonding orbital, and two in the antibonding orbital. The one orbital almost exactly cancels out the effect of the other. Experimentally, the bond energy of dihelium is only 084 kJ/mol. this is not enough to hold the two atoms together in the presence of random thermal motion at ordinary temperatures, so dihelium dissociates as quickly as it is formed, and is therefore not a distinct chemical species.

Diatomic Molecules Containing Second-row Atoms

The four simplest molecules we have examined so far involve molecular orbitals that derived from two 1s atomic orbitals. If we wish to extend our model to larger atoms, we will have to contend with higher atomic orbitals as well. One greatly simplifying principle here is that only the valence-shell orbitals need to be considered. Inner atomic orbitals such as 1s are deep within the atom and well-shielded from the electric field of a neighboring nucleus, so that these orbitals largely retain their atomic character when bonds are formed.

Dilithium

For example, when lithium, whose configuration is $1s^2 2s^1$, bonds with itself to form Li_2, we can forget about the 1s atomic orbitals and consider only the σ bonding and antibonding orbitals. Since there are not enough electrons to populate the antibonding orbital, the attractive forces win out and we have a stable molecule.

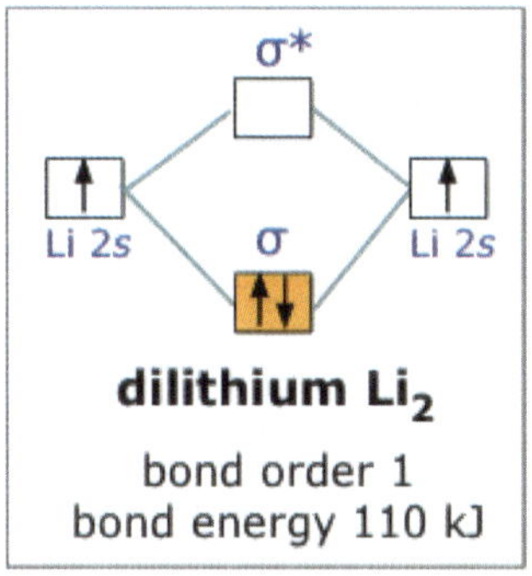

The bond energy of dilithium is 110 kJ/mole; notice that this value is less than half of the 270 kJ bond energy in dihydrogen, which also has two electrons in a bonding orbital. The reason, of course, is that the 2s orbital of Li is much farther from its nucleus than is the 1s orbital of H, and this is equally true for the corresponding molecular orbitals. It is a general rule, then, that the larger the parent atom, the less stable will be the corresponding diatomic molecule.

Lithium Hydride

All the molecules we have considered thus far are homonuclear; they are made up of one kind of atom. As an example of a heteronuclear molecule, let's take a look at a very simple example— lithium hydride. Lithium hydride is a stable, though highly reactive molecule. The diagram shows how the molecular orbitals in lithium hydride can be related to the atomic orbitals of the parent atoms. One thing that makes this diagram look different from the ones we have seen previously is that the parent atomic orbitals have widely differing energies; the greater nuclear charge of lithium reduces the energy of its 1s orbital to a value well below that of the 1s hydrogen orbital.

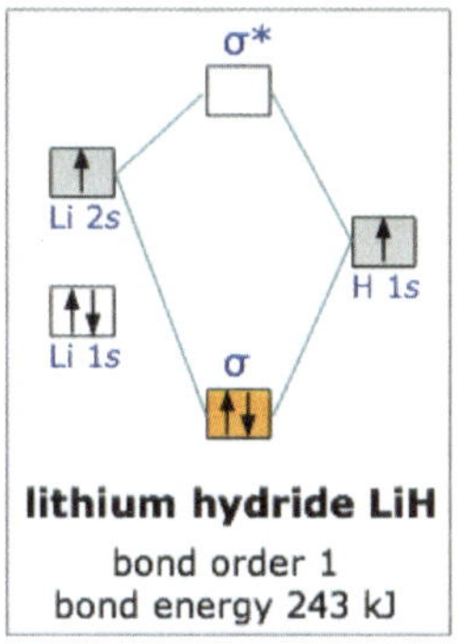

There are two occupied atomic orbitals on the lithium atom, and only one on the hydrogen. With which of the lithium orbitals does the hydrogen 1s orbital interact? The lithium 1s orbital is the lowest-energy orbital on the diagram. Because this orbital is so small and retains its electrons so tightly, it does not contribute to bonding; we need consider only the 2s orbital of lithium which combines with the 1s orbital of hydrogen to form the usual pair of sigma bonding and antibonding orbitals. Of the four electrons in lithium and hydrogen, two are retained in the lithium 1s orbital, and the two remaining ones reside in the σ orbital that constitutes the Li–H covalent bond.

The resulting molecule is 243 kJ/mole more stable than the parent atoms. As we might expect, the bond energy of the heteronuclear molecule is very close to the average of the energies of the corresponding homonuclear molecules. Actually, it turns out that the correct way to make this comparison is to take the geometric mean, rather than the arithmetic mean, of the two bond energies. The geometric mean is simply the square root of the product of the two energies.

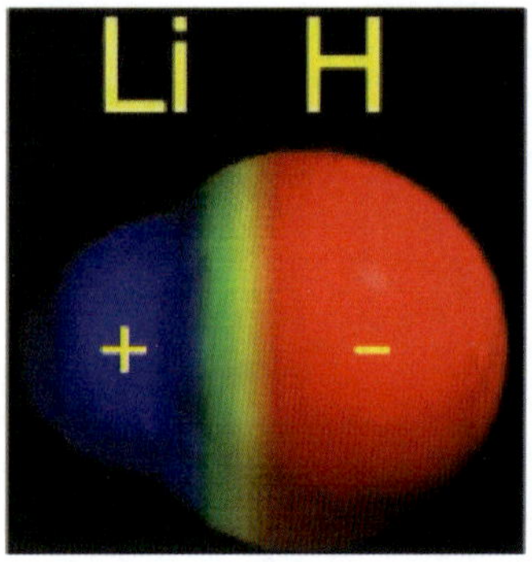

The geometric mean of the H_2 and Li_2 bond energies is 213 kJ/mole, so it appears that the lithium hydride molecule is 30 kJ/mole more stable than it "is supposed" to be. This is attributed to the fact that the electrons in the 2σ bonding orbital are not equally shared between the two nuclei; the orbital is skewed slightly so that the electrons are attracted somewhat more to the hydrogen atom. This bond polarity, arises from the greater electron-attracting power of hydrogen— a consequence of the very small size of this atom. The electrons can be at a lower potential energy if they are slightly closer to the hydrogen end of the lithium hydride molecule. It is worth pointing out, however, that the electrons are, on the average, also closer to the lithium nucleus, compared to where they would be in the 2s orbital of the isolated lithium atom. So it appears that everyone gains and no one loses here.

Σ and π orbitals

The molecules we have considered thus far are composed of atoms that have no more than four electrons each; our molecular orbitals have therefore been derived from s-type atomic orbitals only. If we wish to apply our model to molecules involving larger atoms, we must take a close look at the way in which p-type orbitals interact as well. Although two atomic p orbitals will be expected to split into bonding and antibonding orbitals just as before, it turns out that the extent of this splitting, and thus the relative energies of the resulting molecular orbitals, depend very much on the nature of the particular p orbital that is involved.

There are three possible p orbitals for any value of the principal quantum number. You should also recall that p orbitals are not spherical like orbitals, but are elongated, and thus possess definite directional properties. The three p orbitals correspond to the three directions of Cartesian space, and are frequently designated p_x, p_y, and p_z, to indicate the axis along which the orbital is aligned. Of course, in the free atom, where no coordinate system is defined, all directions are equivalent, and so are the p orbitals. But when the atom is near another atom, the electric field due to that other atom acts as a point of reference that defines a set of directions. The line of centers between the two nuclei is conventionally taken as the x axis. If this direction is represented horizontally on a sheet of paper, then the y axis is in the vertical direction and the z axis would be normal to the page.

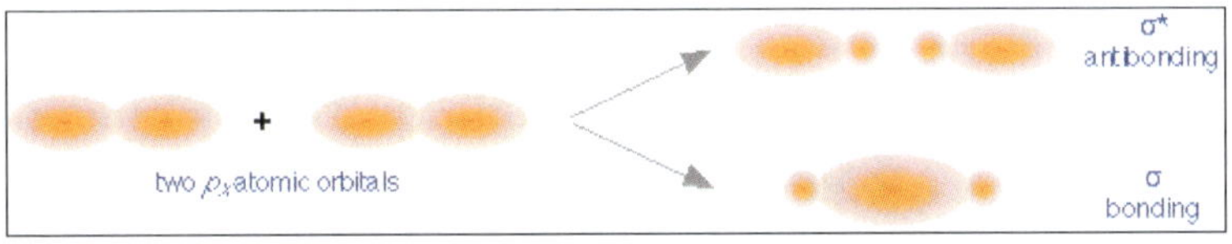

These directional differences lead to the formation of two different classes of molecular orbitals. The above figure shows how two px atomic orbitals interact. In many ways the resulting molecular orbitals are similar to what we got when s atomic orbitals combined; the bonding orbital has a large electron density in the region between the two nuclei, and thus corresponds to the lower potential energy. In the out-of-phase combination, most of the electron density is away from the internuclear region, and as before, there is a surface exactly halfway between the nuclei that corresponds to zero electron density. This is clearly an antibonding orbital— again, in general shape, very much like the kind we saw in hydrogen and similar molecules. Like the ones derived from s-atomic orbitals, these molecular orbitals are σ (sigma) orbitals.

Sigma orbitals are cylindrically symmetric with respect to the line of centers of the nuclei; this means that if you could look down this line of centers, the electron density would be the same in all directions.

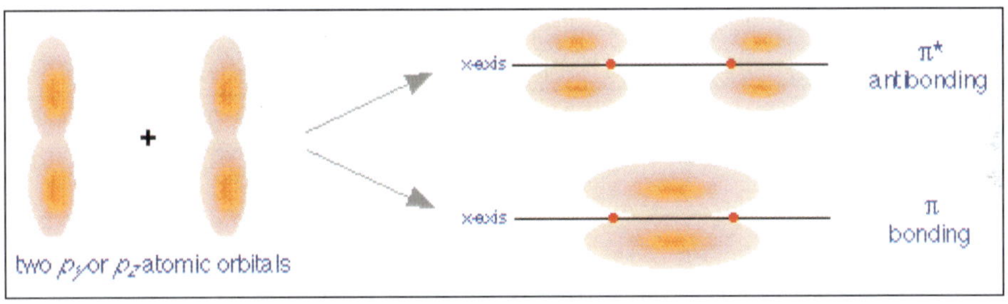

Orbitals, we get the bonding and antibonding pairs that we would expect, but the resulting molecular orbitals have a different symmetry: rather than being rotationally symmetric about the line of centers, these orbitals extend in both perpendicular directions from this line of centers. Orbitals having this more complicated symmetry are called π (pi) orbitals. There are two of them, πy and πz differing only in orientation, but otherwise completely equivalent.

The different geometric properties of the π and σ orbitals causes the latter orbitals to split more than the π orbitals, so that the σ* antibonding orbital always has the highest energy. The σ bonding orbital can be either higher or lower than the π bonding orbitals, depending on the particular atom.

Second-row Diatomics

If we combine the splitting schemes for the 2s and 2p orbitals, we can predict bond order in all of the diatomic molecules and ions composed of elements in the first complete row of the periodic table. Remember that only the valence orbitals of the atoms need be considered; as we saw in the cases of lithium hydride and dilithium, the inner orbitals remain tightly bound and retain their localized atomic character.

Dicarbon

Carbon has four outer-shell electrons, two 2s and two 2p. For two carbon atoms, we therefore have a total of eight electrons, which can be accommodated in the first four molecular orbitals. The lowest two are the 2s-derived bonding and antibonding pair, so the "first" four electrons make no net contribution to bonding. The other four electrons go into the pair of pibonding orbitals, and there are no more electrons for the antibonding orbitals— so we would expect the dicarbon molecule to be stable, and it is. (But being extremely reactive, it is known only in the gas phase).

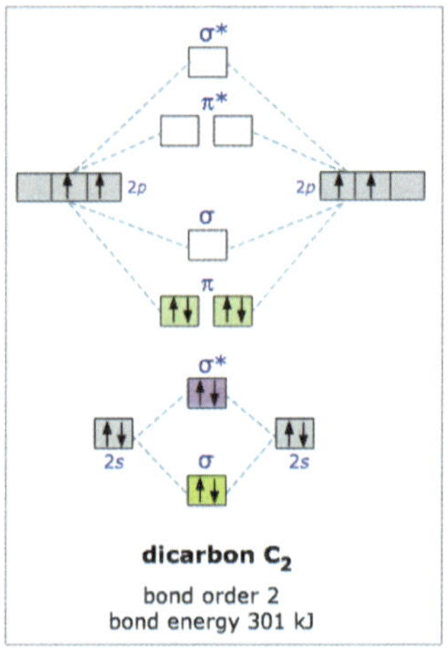

One pair of electrons shared between two atoms constitutes a "single" chemical bond; this is Lewis' original definition of the covalent bond. In C_2 there are two paris of electrons in the π bonding orbitals, so we have what amounts to a double bond here; in other words, the bond order in dicarbon is two.

Dioxygen

The electron configuration of oxygen is $1s^2 2s^2 2p^4$. In O_2, therefore, we need to accommodate twelve valence electrons (six from each oxygen atom) in molecular orbitals. As you can see from the diagram, this places two electrons in antibonding orbitals. Each of these electrons occupies a separate π* orbital because this leads to less electron-electron repulsion (Hund's Rule).

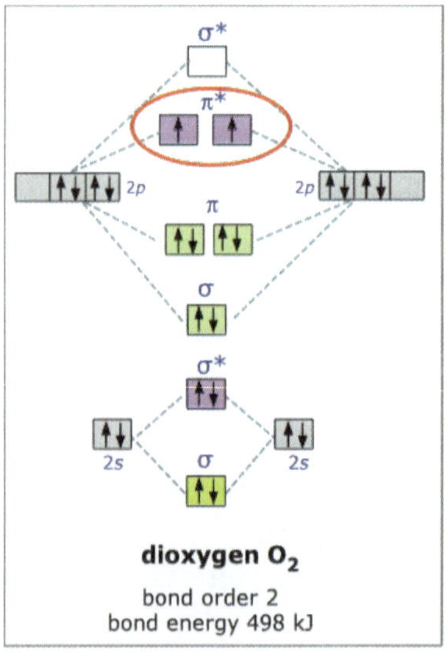

The bond energy of molecular oxygen is 498 kJ/mole. This is smaller than the 945 kJ bond energy of N_2— not surprising, considering that oxygen has two electrons in an antibonding orbital, compared to nitrogen's one.

The two unpaired electrons of the dioxygen molecule give this substance an unusual and distinctive property: O_2 is paramagnetic. The paramagnetism of oxygen can readily be demonstrated by pouring liquid O_2 between the poles of a strong permanent magnet; the liquid stream is trapped by the field and fills up the space between the poles.

Since molecular oxygen contains two electrons in an antibonding orbital, it might be possible to make the molecule more stable by removing one of these electrons, thus increasing the ratio of bonding to antibonding electrons in the molecule. Just as we would expect, and in accord with our model, O_2^+ has a bond energy higher than that of neutral dioxygen; removing the one electron actually gives us a more stable molecule. This constitutes a very good test of our model of bonding and antibonding orbitals. In the same way, adding an electron to O_2 results in a weakening of the bond, as evidenced by the lower bond energy of O_2^-. The bond energy in this ion is not known, but the length of the bond is greater, and this is indicative of a lower bond energy. These two dioxygen ions, by the way, are highly reactive and can be observed only in the gas phase.

VSEPR

The Valence Shell Electron Pair Repulsion Theory abbreviated as VSEPR theory is based on the premise that there is a repulsion between the pairs of valence electrons in all atoms, and the atoms will always tend to arrange themselves in a manner in which this electron pair repulsion is minimalized. This arrangement of the atom determines the geometry of the resulting molecule.

The different geometries that molecules can assume keeping with VSEPR theory can be seen in the illustration provided below.

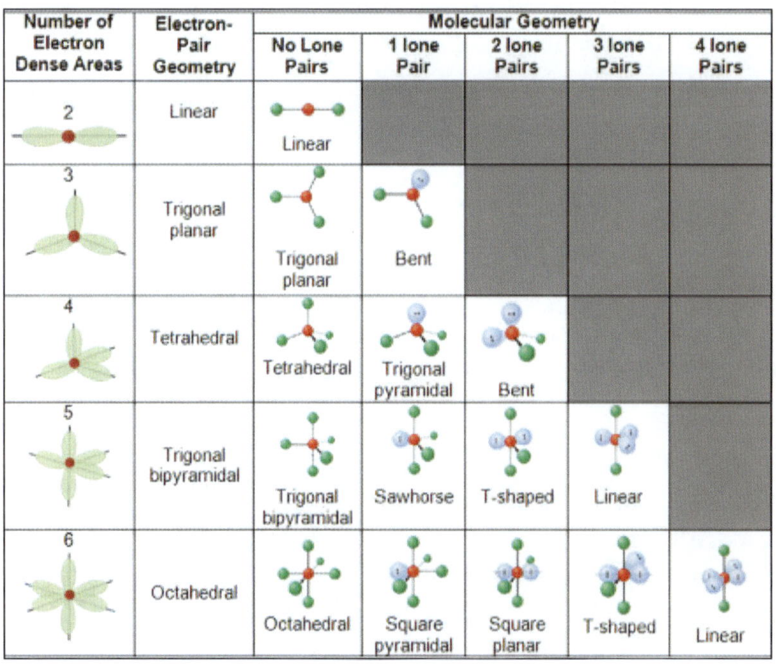

VSEPR Theory – Different geometries that molecules can assume.

The two primary founders of the VSEPR theory are Ronald Nyholm and Ronald Gillespie. This theory is also known as the Gillespie-Nyholm theory to honour these chemists.

Postulates of VSEPR Theory

The postulates of the VSEPR theory are listed below:

- In polyatomic molecules (i.e. molecules made up of three or more atoms), one of the constituent atoms is identified as the central atom to which all other atoms belonging to the molecule are linked.
- The total number of valence shell electron pairs decides the shape of the molecule.
- The electron pairs have a tendency to orient themselves in a way that minimizes the electron-electron repulsion between them and maximizes the distance between them.
- The valence shell can be thought of as a sphere wherein the electron pairs are localized on the surface in such a way that the distance between them is maximized.
- Should the central atom of the molecule be surrounded by bond pairs of electrons, then, the asymmetrically shaped molecule can be expected.
- Should the central atom be surrounded by both lone pairs and bond pairs of electrons, the molecule would tend to have a distorted shape.
- The VSEPR theory can be applied to each resonance structure of a molecule.
- The strength of the repulsion is strongest in two lone pairs and weakest in two bond pairs.
- If electron pairs around the central atom are closer to each other, they will repel each other. This results in an increase in the energy of the molecules.
- If the electron pairs lie far from each other, the repulsions between them will be less and eventually, the energy of the molecule will be low.

Limitations of VSEPR Theory

Some significant limitations of the VSEPR theory include:

- This theory fails to explain isoelectronic species (i.e. elements having the same number of electrons). The species may vary in shapes despite having the same number of electrons.
- The VSEPR theory does not shed any light on the compounds of transition metals. The structure of several such compounds cannot be correctly described by this theory. This is because the VSEPR theory does not take into account the associated sizes of the substituent groups and the lone pairs that are inactive.
- Another limitation of VSEPR theory is that it predicts that halides of group 2 elements will have a linear structure, whereas their actual structure is a bent one.

Predicting the Shapes of Molecules

The following steps must be followed in order to decide the shape of a molecule:

- The least electronegative atom must be selected as the central atom (since this atom has the highest ability to share its electrons with the other atoms belonging to the molecule).
- The total number of electrons belonging to the outermost shell of the central atom must be counted.
- The total number of electrons belonging to other atoms and used in bonds with the central atom must be counted.
- These two values must be added in order to obtain the valence shell electron pair number or the VSEP number.

VSEP Number

The VSEP number describes the shape of the molecule.

VSEP Number	Shape of the Molecule
2	Linear
3	Trigonal Planar
4	Tetrahedral
5	Trigonal Bipyramidal
6	Octahedral
7	Pentagonal Bipyramidal

Each of these corresponding shapes can also be found in the illustration provided earlier. However, the VSEPR theory cannot be used to obtain the exact bond angles between the atoms in a molecule.

Linear Shape of Molecule:

- In this type of molecule, we find two places in the valence shell of the central atom.
- They should be arranged in such a manner such that repulsion can be minimized (pointing in the opposite direction).
- Example: BeF_2.

Trigonal Planar Shape of Molecule:

- In this type of molecule, we find three molecules attached to a central atom.
- They are arranged in such a manner such that repulsion between the electrons can be minimized (toward the corners of an equilateral triangle).
- Example: BF_3.

Tetrahedral Shape of Molecule:

- In two-dimensional molecules, atoms lie in the same plane and if we place these conditions on methane, we will get a square planar geometry in which the bond angle between H-C-H is 90°.

Now, if we consider all these conditions for a three-dimensional molecule, we will get a tetrahedral molecule in which the bond angle between H-C-H is 109°28' (toward the corners of an equilateral triangle) CH_4.

Trigonal Bipyramid Shape of Molecule:

- Let's take an example of PF_5. Here, repulsion can be minimized by even distribution of electrons towards the corner of a trigonal pyramid. In trigonal bipyramid, three positions lie along the equator of the molecule. The two positions lie along an axis perpendicular to the equatorial plane.

Use of VSEPR Theory Predict the Shapes of Molecules

The strength of the repulsion between a lone pair and a bond pair of electrons lies in between the repulsion between two lone pairs and between two bond pairs. The order of repulsion between electron pairs as follows:

Lone Pair - lone pair > Lone Pair - bond - pair > Bond Pair - bond pair.

- Total number of electron pairs around the central atom = ½ (number of valence electrons of central atom + number of atoms linked to central atom by single bonds).
    - For negative ions, add the number of electrons equal to the units of negative charge on the ions to the valence electrons of the central atom.
    - For positive ions, subtract the number of electrons equal to the units of positive charge on the ion from the valence electrons of the central atom.
- The number of Bond pair = Total number of atoms linked to central atom by single bonds.
- Number of lone pairs = Total number of electron – No of shared pair.

The electron pairs around the central atom repel each another and move so far apart from each another that there are no greater repulsions between them. This results in the molecule having minimum energy and maximum stability.

- The shape of a molecule with only two atoms is always linear.
- For molecules with three or more atoms, one of the atoms is called the central atom and other atoms are attached to the central atom.
- If the central atom is linked to similar atoms and is surrounded by bond pairs of electrons only, the repulsions between them are similar as a result the shape of the molecule is symmetrical and the molecule is said to have regular geometry.
- If the central atom is linked to different atoms or is surrounded by bond pair as well as a lone pair of electrons, the repulsion between them is similar. As a result, the shape of the molecule has an irregular or distorted geometry.
- The exact shape of the molecule depends upon the total number of electron pairs present around the central atom.

Hybridization

Hybridization is defined as the concept of mixing two atomic orbitals with the same energy levels to give a degenerated new type of orbitals. This intermixing is based on quantum mechanics. The atomic orbitals of the same energy level can only take part in hybridization and both full filled and half-filled orbitals can also take part in this process provided they have equal energy.

During the process of hybridization, the atomic orbitals of similar energy are mixed together such as the mixing of two 's' orbitals or two 'p' orbital's or mixing of an 's' orbital with a 'p' orbital or 's' orbital with a 'd' orbital.

Redistribution of the energy of orbitals of individual atoms to give orbitals of equivalent energy happens when two atomic orbitals combine together to form hybrid orbital in a molecule. This process is called hybridization. The new orbitals thus formed are known as hybrid orbitals.

Types of Hybridization

Based on the types of orbitals involved in mixing, the hybridization can be classified as sp3, sp2, sp, sp3d, sp3d2, sp3d3.

sp Hybridization

sp hybridization is observed when one s and one p orbital in the same main shell of an atom mix to form two new equivalent orbitals. The new orbitals formed are called sp hybridized orbitals. It forms linear molecules with an angle of 180°:

- This type of hybridization involves the mixing of one 's' orbital and one 'p' orbital of equal energy to give a new hybrid orbital known as an sp hybridized orbital.
- sp hybridization is also called diagonal hybridization.
- Each sp hybridized orbital has an equal amount of s and p character, i.e., 50% s and p character.

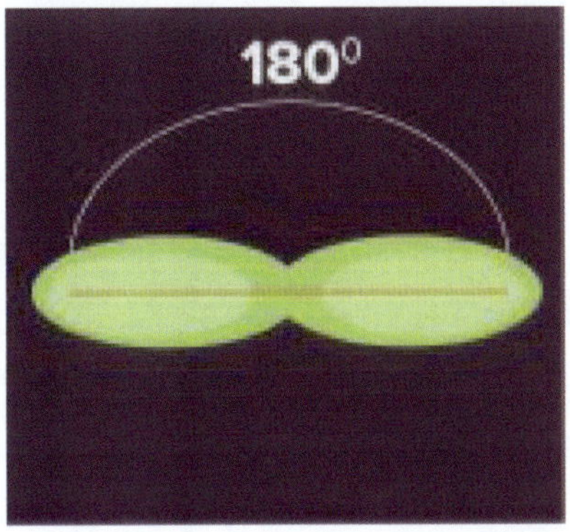

Examples of sp Hybridization:

- All compounds of beryllium like BeF_2, BeH_2, $BeCl_2$.
- All compounds of carbon-containing triple Bond like C_2H_2.

sp^2 Hybridization

- sp^2 hybridization is observed when one s and two p orbitals of the same shell of an atom mix to form 3 equivalent orbital. The new orbitals formed are called sp^2 hybrid orbitals.
- sp^2 hybridization is also called trigonal hybridization.
- It involves mixing of one 's' orbital and two 'p' orbital's of equal energy to give a new hybrid orbital known as sp^2.
- A mixture of s and p orbital formed in trigonal symmetry and is maintained at 120°.
- All the three hybrid orbitals remain in one plane and make an angle of 120° with one another. Each of the hybrid orbitals formed has 33.33% s character and 66.66% 'p' character.
- The molecules in which the central atom is linked to 3 atoms and is sp^2 hybridized have a triangular planar shape.

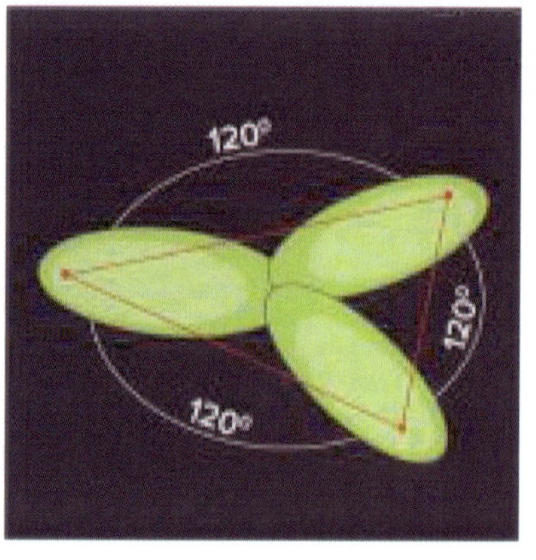

Examples of sp^2 Hybridization:

- All the compounds of Boron i.e. BF_3, BH_3.
- All the compounds of carbon containing a carbon-carbon double bond, Ethylene (C_2H_4).

sp^3 Hybridization

When one 's' orbital and 3 'p' orbitals belonging to the same shell of an atom mix together to form four new equivalent orbital, the type of hybridization is called a tetrahedral hybridization or sp^3. The new orbitals formed are called sp^3 hybrid orbitals.

- These are directed towards the four corners of a regular tetrahedron and make an angle of 109°28' with one another.
- The angle between the sp^3 hybrid orbitals is 109.28°.
- Each sp^3 hybrid orbital has 25% s character and 75% p character.
- Example of sp^3 hybridization: ethane (C_2H_6), methane.

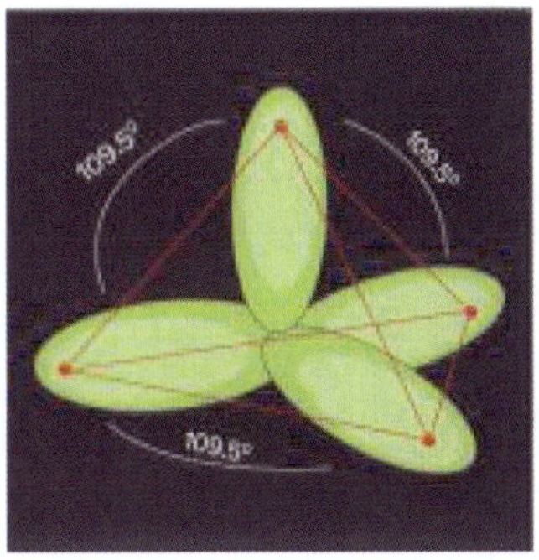

sp³d Hybridization

sp³d hybridization involves the mixing of 3p orbitals and 1d orbital to form 5 sp3d hybridized orbitals of equal energy. They have trigonal bipyramidal geometry.

- The mixture of s, p and d orbital forms trigonal bipyramidal symmetry.
- Three hybrid orbitals lie in the horizontal plane inclined at an angle of 120° to each other known as the equatorial orbitals.
- The remaining two orbitals lie in the vertical plane at 90 degrees plane of the equatorial orbitals known as axial orbitals.
- Example: Hybridization in Phosphorus pentachloride (PCl_5).

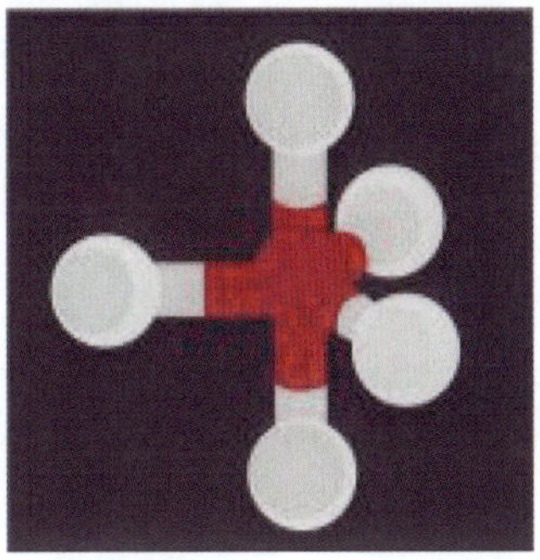

sp³d² Hybridization

- sp³d² hybridization has 1s, 3p and 2d orbitals, that undergo intermixing to form 6 identical sp³d² hybrid orbitals.
- These 6 orbitals are directed towards the corners of an octahedron.
- They are inclined at an angle of 90 degrees to one another.

Bent's Rule

Bent's rule describes and explains the relationship between the orbital hybridization of central atoms in molecules and the electronegativities of substituents. The rule was stated by Henry Bent as follows: "Atomic character concentrates in orbitals directed toward electropositive substituents".

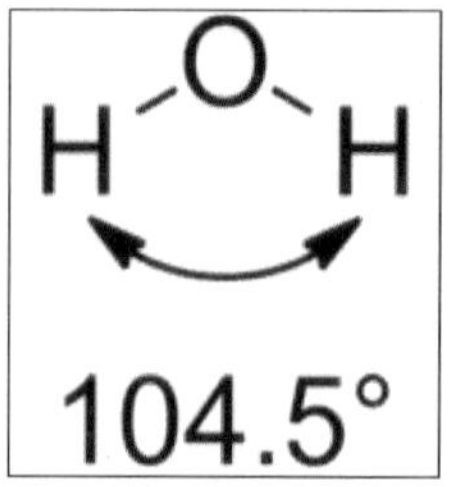

Structure of water illustrating how the bond angle deviates from the tetrahedral angle of 109.5°.

The chemical structure of a molecule is intimately related to its properties and reactivity. Valence bond theory proposes that molecular structures are due to covalent bonds between the atoms and that each bond consists of two overlapping and typically hybridised atomic orbitals. Traditionally, p-block elements in molecules are assumed to hybridise strictly as sp^n, where n is either 1, 2, or 3. In addition, the hybrid orbitals are all assumed to be equivalent (i.e. the n+1 sp^n orbitals have the same p character). Results from this approach are usually good, but they can be improved upon by allowing isovalent hybridization, in which the hybridised orbitals may have noninteger and unequal p character. Bent's rule provides a qualitative estimate as to how these hybridised orbitals should be constructed. Bent's rule is that in a molecule, a central atom bonded to multiple groups will hybridise so that orbitals with more s character are directed towards electropositive groups, while orbitals with more p character will be directed towards groups that are more electronegative. By removing the assumption that all hybrid orbitals are equivalent sp^n orbitals, better predictions and explanations of properties such as molecular geometry and bond strength can be obtained.

Bent's rule can be generalized to d-block elements as well. The hybridisation of a metal center is arranged so that orbitals with more s character are directed towards ligands that form bonds with more covalent character. Equivalently, orbitals with more d character are directed towards groups that form bonds of greater ionic character.

Bent's rule can be used to explain trends in both molecular structure and reactivity. After determining how the hybridisation of the central atom should affect a particular property, the electronegativity of substituents can be examined to see if Bent's rule holds.

Bond Angles

Knowing the angles between bonds is a crucial component in determining a molecular structure. In valence bond theory, covalent bonds are assumed to consist of two electrons lying in overlapping, usually hybridised, atomic orbitals from bonding atoms. Orbital hybridisation explains why methane is tetrahedral and ethylene is planar for instance. However, there are deviations from the ideal geometries of sp^n hybridisation such as in water and ammonia. The bond angles in those molecules are 104.5° and 107° respectively, which are below the expected tetrahedral angle of 109.5°. The traditional approach to explain those differences is VSEPR theory. In that framework, valence electrons are assumed to lie in localized regions and lone pairs are assumed to repel each other to a greater extent than bonding pairs.

Bent's rule provides an alternative explanation as to why some bond angles differ from the ideal geometry. First, a trend between central atom hybridisation and bond angle can be determined by using the model compounds methane, ethylene, and acetylene. In order, the carbon atoms are directing sp^3, sp^2, and sp orbitals towards the hydrogen substituents. The bond angles between substituents are 109.5°, ~120°, and 180°. This simple system demonstrates that hybridised atomic orbitals with higher p character will have a smaller angle between them. This result can be made rigorous and quantitative as Coulson's theorem.

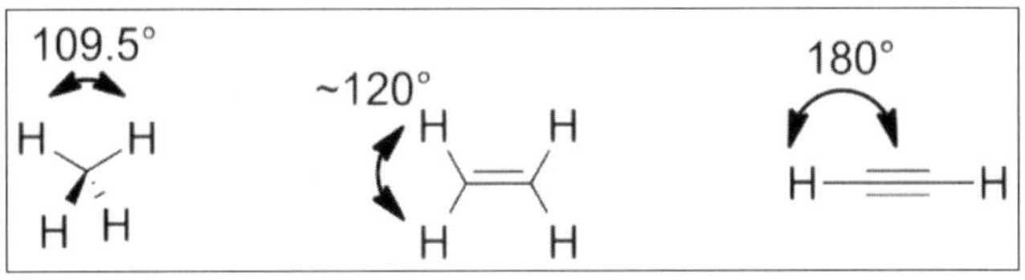

Now that the connection between hybridisation and bond angles has been made, Bent's rule can be applied to specific examples. The following were used in Bent's original paper, which considers the group electronegativity of the methyl group to be less than that of the hydrogen atom because methyl substitution reduces the acid dissociation constants of formic acid and of acetic acid.

Molecule	Bond angle between substituents
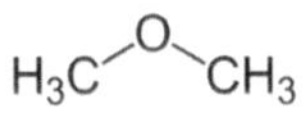 Dimethyl ether	111°
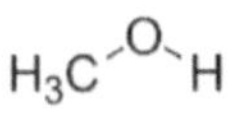 Methanol	107-109°
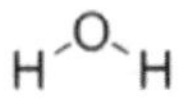 Water	104.5°
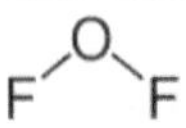 Oxygen difluoride	103.8°

As one moves down the table, the substituents become more electronegative and the bond angle between them decreases. According to Bent's rule, as the substituent electronegativies increase, orbitals of greater p character will be directed towards those groups. This will decrease the bond angle. This agrees with the experimental results. Comparing this explanation with VSEPR theory, VSEPR cannot explain why the angle in dimethyl ether is greater than 109.5°.

In predicting the bond angle of water, Bent's rule suggests that hybrid orbitals with more s character should be directed towards the very electropositive lone pairs, while that leaves orbitals with more p character directed towards the hydrogens. This increased p character in those orbitals decreases the bond angle between them to less than the tetrahedral 109.5°. The same logic can be applied to ammonia, the other canonical example of this phenomenon.

Bond Lengths

Similarly to bond angles, the hybridisation of an atom can be related to the lengths of the bonds it forms. As bonding orbitals increase in s character, the σ bond length decreases.

Molecule	Average carbon–carbon bond length
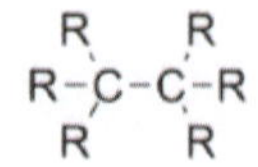	1.54 Å
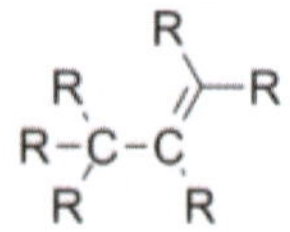	1.50 Å
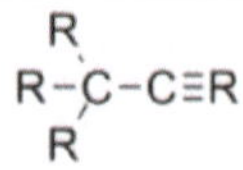	1.46 Å

By adding electronegative substituents and changing the hybridisation of the central atoms, bond lengths can be manipulated. If a molecule contains a structure X-A--Y, replacement of the substituent X by a more electronegative atom changes the hybridization of central atom A and shortens the adjacent A--Y bond.

Molecule	Average carbon–fluorine bond length
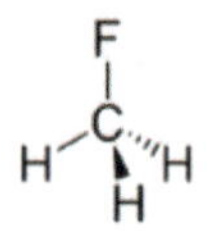 Fluoromethane	1.388 Å
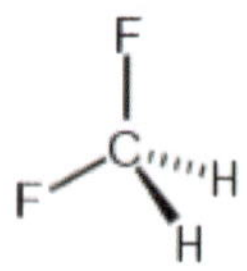 Difluoromethane	1.358 Å
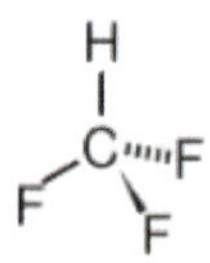 Trifluoromethane	1.329 Å
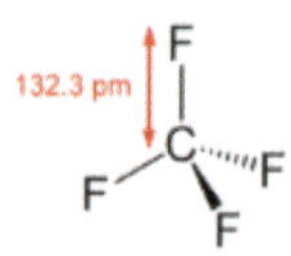 Tetrafluorometh-ane	1.323 Å

Because fluorine is so much more electronegative than hydrogen, in fluoromethane the carbon will direct hybrid orbitals higher in s character towards the three hydrogens than towards the fluorine. In difluoromethane, there are only two hydrogens so less s character in total is directed towards them and more is directed towards the two fluorines, which shortens the C—F bond lengths relative to fluoromethane. This trend holds all the way to tetrafluoromethane whose C-F bonds have the highest s character (25%) and the shortest bond lengths in the series.

The same trend also holds for the chlorinated analogs of methane, although the effect is less dramatic because chlorine is less electronegative than fluorine.

Molecule	Average carbon–chlorine bond length
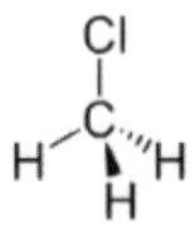 Chloromethane	1.783 Å
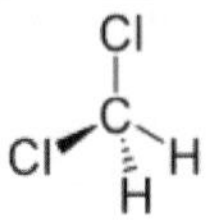 Dichloromethane	1.772 Å
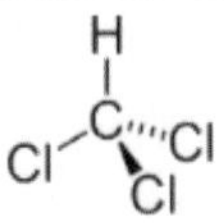 Trichloromethane	1.767 Å
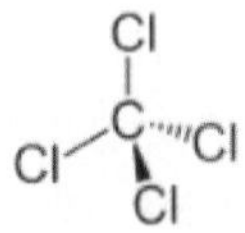 Tetrachloromethane	1.766 Å

The above cases seem to demonstrate that the size of the chlorine is less important than its electronegativity. A prediction based on sterics alone would lead to the opposite trend, as the large chlorine substituents would be more favorable far apart. As the steric explanation contradicts the experimental result, Bent's rule is likely playing a primary role in structure determination.

J_{CH} Coupling Constants

Perhaps the most direct measurement of s character in a bonding orbital between hydrogen and carbon is via the ^{1}H–^{13}C coupling constants determined from NMR spectra. Theory predicts that J_{CH} values will be much higher in bonds with more s character.

Molecule	J_{CH} (of the methyl protons)
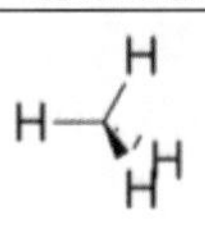 Methane	125 Hz
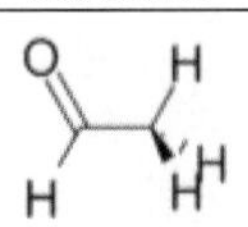 Acetaldehyde	127 Hz
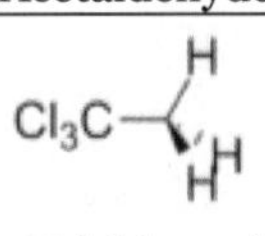 1,1,1–Trichloroethane	134 Hz
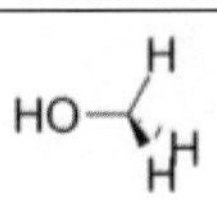 Methanol	141 Hz

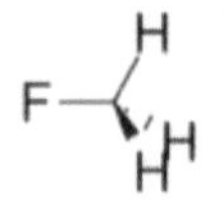 Fluoromethane	149 Hz

As the electronegativity of the substituent increases, the amount of p character directed towards the substituent increases as well. This leaves more s character in the bonds to the methyl protons, which leads to increased J_{CH} coupling constants.

Inductive Effect

The inductive effect can be explained with Bent's rule. The inductive effect is the transmission of charge through covalent bonds and Bent's rule provides a mechanism for such results via differences in hybridisation. As the groups bonded to the central carbon become more electronegative, the central carbon becomes more electron-withdrawing as measured by the polar substituent constant. The polar substituent constants are similar in principle to σ values from the Hammett equation, as an increasing value corresponds to a greater electron-withdrawing ability. Bent's rule suggests that as the electronegativity of the groups increase, more p character is diverted towards those groups, which leaves more s character in the bond between the central carbon and the R group. As s orbitals have greater electron density closer to the nucleus than p orbitals, the electron density in the C–R bond will more shift towards the carbon as the s character increases. This will make the central carbon more electron-withdrawing to the R group. Thus, the electron-withdrawing ability of the substituents has been transferred to the adjacent carbon, exactly what the inductive effect predicts.

Substituent	Polar substituent constant (larger values imply greater electron-withdrawing ability)
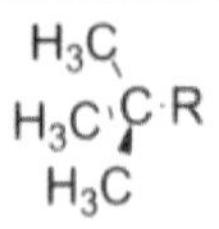 t–Butyl	-0.30
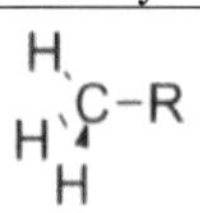 Methyl	0.00
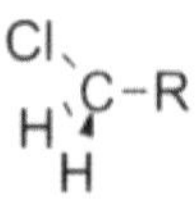 Chloromethyl	1.05
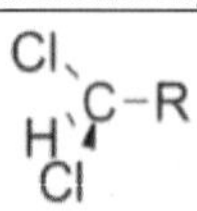 Dichloromethyl	1.94
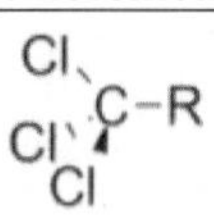 Trichloromethyl	2.65

References

- Unitcell: chem.purdue.edu, Retrieved 4 February, 2019
- Imperfections-or-defects-in-a-solid, the-solid-state, chemistry: toppr.com, Retrieved 16 May, 2019
- Atomic-structure: byjus.com, Retrieved 22 August, 2019
- Definition-of-valence-bond-theory: thoughtco.com, Retrieved 11 February, 2019
- Valence-bond-theory, chemistry: byjus.com, Retrieved 9 April, 2019
- Valence-bond-theory, boundless-chemistry: lumenlearning.com, Retrieved 19 May, 2019
- Molecular-orbital-theory, chemical-bonding-and-molecular-structure, chemistry: toppr.com, Retrieved 13 January, 2019
- A-Molecular-Orbital-Theory, Chemical-Bonding-and-Molecular-Structure, Chemistry: libretexts.org, Retrieved 23 June, 2019
- Postulates-of-VSEPR-Theory, vsepr-theory: byjus.com, Retrieved 21 January, 2019
- Hybridization: byjus.com, Retrieved 8 March, 2019
- Bent's-rule: ipfs.io, Retrieved 17 July, 2019

3

Understanding Inorganic Compounds

A compound which does not have a carbon to hydrogen bond is known as an inorganic compound. Some of the common inorganic compounds are water, halogens and halides and metal hydrides. The topics elaborated in this chapter will help in gaining a better perspective about the properties of these inorganic compounds.

An inorganic compound can be considered as a compound that does not contain a carbon-to-hydrogen bond, also called a C-H bond. Moreover, inorganic compounds tend to be minerals or geologically-based compounds that do not contain carbon-to-hydrogen bonds. Not all, but most inorganic compounds contain a metal. That said, there are countless compounds that fall under the realm of inorganic. In fact, the majority of all compounds in this universe are inorganic in nature. For this reason, inorganic compounds have an overwhelming amount of applications and practical uses in the real world. Since most of the compounds in this world are inorganic, these compounds can take on a host of forms and possess many different characteristics.

Characteristics

Since many inorganic compounds contain some type of metal (alkali, alkaline, transition, etc.), they tend to be able to conduct electricity. For example, while in the solid state, inorganic compounds are poor conductors of electricity. However, in the liquid phase, inorganic compounds are highly conductive. In this phase, inorganic compounds' electrons are able to move very freely, and this movement of electrons is noted as electricity.

Due to ionic bonding typically found in inorganic compounds, they are held together very rigidly and possess extremely high melting and boiling points. Another distinct characteristic of inorganic compounds is their color. Transition metal inorganic compounds, even sitting on a bench-top, are usually highly colored, and this is, again, due to the configuration of the 'd-block' electrons. The bright and beautiful colors that one sees when fireworks explode is due to the inorganic metal (usually an alkali or alkaline one) present in the compound. Because inorganic compounds display a unique color when burned, this can be used as a 'marker' to identify the metal involved. Also, inorganic compounds are typically highly soluble in water. That is to say, they can 'disappear' when placed into water since they will simply dissolve. Yet another revealing characteristic of inorganic compounds is their ability to form crystals. The nature of the bonding found in inorganic compounds end them to be able to grow crystals in saturated solutions.

There are three groups of inorganic compounds essential to life: water, salts and acids.

Water

Water is the most abundant substance in the body. Its abundance is due partly to its unique chemical properties created by the influence of its hydrogen bonds.

Water as a Lubricant and Cushion

Water is a major component of many of the body's lubricating fluids. Just as oil lubricates the hinge on a door, water in synovial fluid lubricates the actions of body joints, and water in pleural fluid helps the lungs expand and recoil with breathing. Watery fluids help keep food flowing through the digestive tract, and ensure that the movement of adjacent abdominal organs is friction free.

Water also protects cells and organs from physical trauma, cushioning the brain within the skull, for example, and protecting the delicate nerve tissue of the eyes. Water cushions a developing fetus in the mother's womb as well.

Water as a Mechanism for Maintaining Body Temperature

Water absorbs and dissipates heat but does not experience a corresponding increase in temperature. In the body, water absorbs the heat generated by chemical reactions without greatly increasing in temperature. Moreover, when the environmental temperature soars, the water stored in the body helps keep the body cool. This cooling effect happens as warm blood from the body's core flows to the blood vessels just under the skin and is transferred to the environment. At the same time, sweat glands release warm water in sweat. As the water evaporates into the air, it carries away heat, and then the cooler blood from the periphery circulates back to the body core.

Water as a Component of Liquid Mixtures

A mixture is a combination of two or more substances, each of which maintains its own chemical identity. In other words, the constituent substances are not chemically bonded into a new, larger chemical compound. The concept is easy to imagine if you think of powdery substances such as flour and sugar; when you stir them together in a bowl, they obviously do not bond to form a new compound. The room air you breathe is a gaseous mixture, containing three discrete elements—nitrogen, oxygen, and argon—and one compound, carbon dioxide. There are three types of liquid mixtures, all of which contain water as a key component. These are solutions, colloids, and suspensions.

For cells in the body to survive, they must be kept moist in a water-based liquid called a solution. In chemistry, a liquid solution consists of a solvent that dissolves a substance called a solute. An important characteristic of solutions is that they are homogeneous; that is, the solute molecules are distributed evenly throughout the solution. If you were to stir a teaspoon of sugar into a glass of water, the sugar would dissolve into sugar molecules separated by water molecules. The ratio of sugar to water in the left side of the glass would be the same as the ratio of sugar to water in the right side of the glass. If you were to add more sugar, the ratio of sugar to water would change, but the distribution—provided you had stirred well—would still be even.

Water is considered the "universal solvent" and it is believed that life cannot exist without water because of this. Water is certainly the most abundant solvent in the body; essentially all of the body's chemical reactions occur among compounds dissolved in water. Because water molecules are polar, with regions of positive and negative electrical charge, water readily dissolves ionic compounds and polar covalent compounds. Such compounds are referred to as hydrophilic, or "water-loving." Sugar dissolves well in water. This is because sugar molecules contain regions of hydrogen-oxygen polar bonds, making it hydrophilic. Nonpolar molecules, which do not readily dissolve in water, are called hydrophobic, or "water-fearing."

Role of Water in Chemical Reactions

Two types of chemical reactions involve the creation or the consumption of water: dehydration synthesis and hydrolysis.

- In dehydration synthesis, one reactant gives up an atom of hydrogen and another reactant gives up a hydroxyl group (OH) in the synthesis of a new product. In the formation of their covalent bond, a molecule of water is released as a byproduct. This is also sometimes referred to as a condensation reaction.
- In hydrolysis, a molecule of water disrupts a compound, breaking its bonds. The water is itself split into H and OH. One portion of the severed compound then bonds with the hydrogen atom, and the other portion bonds with the hydroxyl group.

These reactions are reversible, and play an important role in the chemistry of organic compounds.

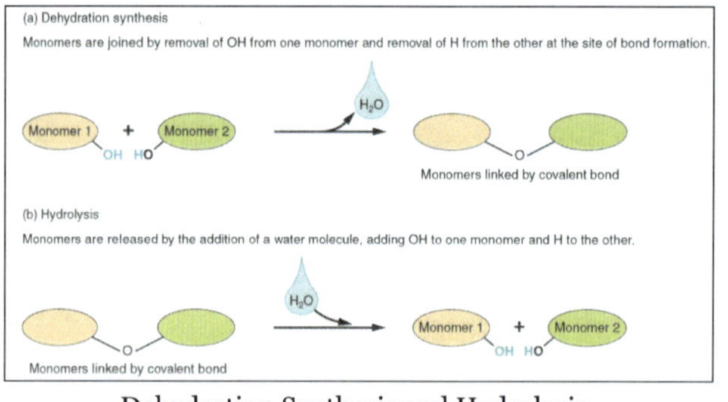

Dehydration Synthesis and Hydrolysis.

Monomers, the basic units for building larger molecules, form polymers (two or more chemically-bonded monomers). (a) In dehydration synthesis, two monomers are covalently bonded in a reaction in which one gives up a hydroxyl group and the other a hydrogen atom. A molecule of water is released as a byproduct during dehydration reactions. (b) In hydrolysis, the covalent bond between two monomers is split by the addition of a hydrogen atom to one and a hydroxyl group to the other, which requires the contribution of one molecule of water.

Acids

Acids are a molecule or other species which can donate a proton or accept an electron pair in reactions. The word "acid" is derived from the Latin word "acids" which means sour. All acid elements have a few things in common i.e. all are sour in taste, they turn blue litmus paper to red, and lose their acidity if they're combined with alkaline substances. The pH level of acids ranges from 0 – 6.

Some common examples of acids are Citrus fruits such as lemons, limes, oranges, grapefruit, etc. All these fruits contain citric acid. Hence, they taste sour or tart. Citric acid is a weak acid but still, it produces hydrogen ions when mixed with water and that's why the pH of lemon juice is 2. Another example of an acid is vinegar. Vinegar consists of acetic acid. Ever wondered why your skin becomes red and swollen after an ant bite or a mosquito bite? Because these insects inject formic acid which causes such skin reactions. Nitric acid, sulfuric acid, hydrochloric acid, etc. are other common ones.

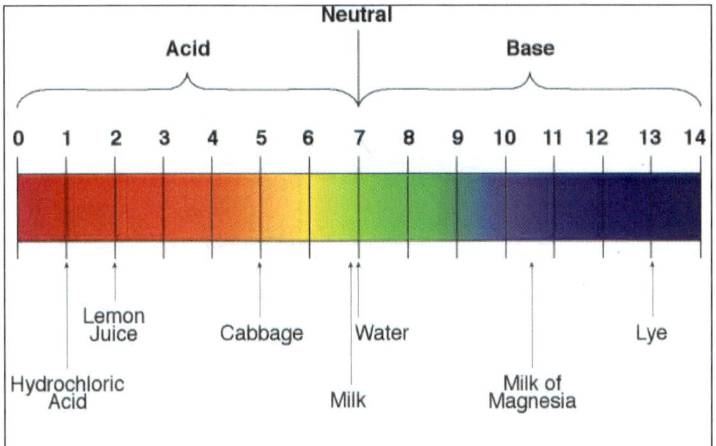

The pH Scale

Properties of Acids

The properties of Acids are as follows:

- Acids change the color of blue litmus to red.
- They change the color of Methyl Orange/Yellow to Pink.
- Acidic substances converts Phenolphthalein from deep pink to colorless.
- Are sour or tart in taste.
- The pH level of acids range from 0-6.
- Acids lose their acidity when combined with alkaline.
- They destroy the chemical property of bases.
- When reacting with metals they produce hydrogen gas.
- Acids produce carbon dioxide when reacted with carbonates.
- Most acids are corrosive in nature which means that they tend to corrode or rust metals.

Classification of Acids

Acids are often classified on the basis of source, the presence of oxygen, strength, concentration and basicity.

1. Classification based on the source:

This means that the acid is classified on the basis of their source or origin. They are mainly of two types: Organic acid and Mineral acid.

- Organic Acid: This is the acid obtained from organic materials such as plants and animals. For e.g. citric acid (Citrus fruits), Acetic acid (Vinegar), Oleic acid (Olive oil), etc.
- Mineral Acid: Mineral acid is procured from minerals. They are also known as inorganic acids. They do not contain carbon. For e.g. H_2SO_4, HCl. HNO_3, etc.

2. Classification based on the presence of Oxygen:

This means that the acids are classified on the basis of the presence of oxygen. These are of two types: Oxy-acid and Hydracids.

- Oxy-acid: Acids that consists oxygen in their composition is known as Oxy-acids. For e.g. H_2SO_4, HNO_3, etc.
- Hydracid: Those that consists hydrogen combined with other elements and do not contain any oxygen in their composition and do not contain any oxygen in their composition are called Hydracids. For e.g. HCl, HI, HBr, etc.

3. Classification based on the Strength of the acid:

Acids produce hydrogen ions when mixed with H_2O, the strength of an acid depends on its concentration of the hydrogen ions present in a solution. A greater number of hydrogen ions means greater strength of the acid whereas, lower number of hydrogen ions means that the acid is weak. They are classified as:

- Strong Acids: An acid which can be dissociated completely or almost completely in water is known as a strong acid. For e.g. sulphuric acid, nitric acid, hydrochloric acid, etc.

  $H^+ + H_2O <\!\!--\!\!> H_3O^+$

  $HCl(aq) <\!\!--\!\!> H^+(aq) + SO_4^-(aq)$

  $H_2SO_4(aq) <\!\!--\!\!> 2H^+(aq) + SO_4^-(aq)$

- Weak Acids: An acid which doesn't dissociate completely or dissociates negligibly in water is known as a weak acid. For e.g. those that which we usually consume on daily basis i.e. citric acid, acetic acid, etc.

  $CH_3COOH_9(aq) <\!\!--\!\!> CH_3COO^-(aq) + H^+(aq)$

  $HCOOH(aq) <\!\!--\!\!> HCOO^-(aq) + H^+(aq)$

4. Classification based on its concentration:

The concentration of the acid depends on the number of hydrogen ions that it produces in water. Based on this the acid is classified as:

- Concentrated Acid: When an aqueous solution has a relatively high percentage of acid dissolved in it, then it is a concentrated acid. For e.g. concentrated hydrochloric acid, concentrated sulphuric acid, concentrated nitric acid, etc.
- Diluted Acid: When an aqueous solution has a relatively low percentage of acid dissolved in it, then it is a dilute acid. For e.g. dilute hydrochloric acid, dilute sulphuric acid, dilute nitric acid, etc.

5. Classification based on the basicity of the acid:

Acid on dissociation in water produces hydrogen ion. The number of these hydrogen ions that can be replaced in an acid is the basicity of an acid.

- Monobasic Acid: A monobasic acid is an acid which has only one hydrogen ion. Therefore, these acids combine with one hydroxyl group of the base to form salt and water. For e.g. HCl, HCOOH, HBr etc.

- Dibasic Acid: Dibasic acid is that which shares two hydroxyl groups it is known as dibasic acid. Dibasic acid dissociates in 2 steps. They can provide 2 kinds of salts i.e. the normal salt and a hydrogen salt:

  $H_2SO_4(aq) \longleftrightarrow H^+(aq) + HSO^-_4(aq)$

  $2NaOH(aq) + H_2SO_4(aq) \longleftrightarrow Na_2SO_4(aq) + 2H_2O(l)$

- Tribasic Acid: Tribasic acids are those which can combine with three hydroxyl groups. They have three replaceable hydrogen ions, and they produce 3 types of salts. For e.g. H_3PO_4:

  $NaOH(aq) + H_3PO_4(aq) \longleftrightarrow NaH_2PO_4(aq) + H_2O(l)$

  $2NaOH(aq) + H_3PO_4(aq) \longleftrightarrow Na_2HPO_4(aq) + 2H_2O(l)$.

Arrhenius, Brønsted-Lowry and Lewis Acids

There are different ways of defining acids. When a person refers to "an acid," this usually refers to an Arrhenius or Brønsted-Lowry acid. A Lewis acid is typically called a "Lewis acid." The reason for the differing definitions is because these different acids don't include the same set of molecules.

- Arrhenius Acid: By this definition, an acid is a substance which increases the concentration of hydronium ions (H_3O^+) when added to water. You may also consider increasing the concentration of hydrogen ion (H^+), as an alternative.

- Brønsted-Lowry Acid: By this definition, an acid is a material capable of acting as a proton donor. This is a less restrictive definition because solvents besides water are not excluded. Essentially, any compound that can be deprotonated is a Brønsted-Lowry acid, including typical acids, plus amines, and alcohol. This is the most widely used definition of an acid.

- Lewis Acid: A Lewis acid is a compound which can accept an electron pair to form a covalent bond. By this definition, some compounds which don't contain hydrogen qualify as acids, including aluminum trichloride and boron trifluoride.

Acid Examples: These are examples of types of acids and specific acids:

- Arrhenius acid
- Monoprotic acid
- Lewis acid
- Hydrochloric acid

- Sulfuric acid
- Hydrofluoric acid
- Acetic acid
- Stomach acid (which contains hydrochloric acid)
- Vinegar (which contains acetic acid)
- Citric acid (found in citrus fruits)

Bases

Bases are substances that slippery to touch when in aqueous form. They taste bitter and change the color of red litmus paper to blue. Bases also dissociate in the water like acids, but instead of producing H^+ they produce OH^- i.e. hydroxyl ion. If a base dissolves in water then it is called an Alkali. Ammonium hydroxide, Calcium hydroxide are some examples of alkalis. Alkalines become less alkaline when mixed with acids. The pH level of bases ranges from 8-14.

Some examples are caustic soda or sodium hydroxide, calcium hydroxide or limewater, borax. A lot of bleaches, soaps, detergents, kinds of toothpaste, etc. are bases.

Properties of Bases

Some of their properties are:

- Bases are slippery to touch when in aqueous form.
- A base will usually taste bitter.
- The pH level of a base is from 8 to 14.
- Bases react with acid to form salt and water.
- A Base will turn red litmus to blue.

Classification of Bases

They are usually classified on the basis of strength, concentration and on its acidity.

1. Classification based on the Strength:

Just like acids, the strength of bases depends on the number of hydroxyl ions it produces when dissolved in water. A high amount of hydroxyl ion represents a strong base and a low amount of base represent a weak base.

- Strong base: A base that dissolves completely or almost completely in water is known as a strong base. For e.g. NaOH, KOH, $Ca(OH)_2$, etc.

  NaOH <—> Na^+ + OH^-

- Weak base: A base which doesn't dissolve completely is called a weak base. For e.g. $Ma(OH)_2$, NH_4OH.

  NH_4OH(aq) <—> NH^+_4(aq) + OH^-(aq)

2. Classification based on the Concentration:

The concentration of the base depends upon the amount of base dissolved in water. It is of two types i.e. Concentrated and Dilute base.

- Concentrated Base: An aqueous solution which has a relatively high percentage of the base is a concentrated base. For e.g. concentrated sodium hydroxide, concentrated potassium hydroxide, concentrated ammonium hydroxide, etc.
- Diluted Base: An aqueous solution which has a relatively low percentage of the base is a dilute base. For e.g. dilute sodium hydroxide, diluted potassium hydroxide, dilute ammonium hydroxide, etc.

3. Classification based on the Acidity of the Base:

The acidity of a base depends on the number of hydroxyl ions it contains. It also depends on the number of hydrogen ions with which a base can combine as one hydrogen ion combines with one hydroxyl ion. It is usually of three type Monoacidic base, Diacidic base and Triacidic base.

- Monoacidic base: It is a base that consists only one hydroxyl ion and only combines with one hydrogen ion. For e.g. NaOH, KOH, NH_4OH, etc.

  NaOH(aq) + HCl(aq) <—> NaCl(aq) + H_2O(l)

- Diacidic base: It is a base that contains two hydroxyl ions and combines with two hydrogen ions. For e.g. $Mg(OH)_2$, $Fe(OH)_2$, $Zn(OH)_2$, etc.

  $Ca(OH)_2$(aq) + 2HCl(aq) <—> $CaCl_2$(aq) + 2H_2O(l)

- Triacidic base: It is a base that has three hydroxyl ions and combines with three hydrogen ions. For e.g. Aluminium Hydroxide.

  $Al(OH)_3$(aq) + 3HCl(aq) <—> $AlCl_3$(aq) + 3H_2O(l)

Salts

Salts are ionic compounds that contain two groups of positively charged ions (H^+) and negatively charged ions (OH^-). In chemistry, salt is produced when acids and bases are combined together in equal proportions. Whereas naturally, salt is found in sea water. The process where acids and bases neutralize each other to form salt and water is known as neutralization. The chemical formula of common salt is NaCl. NaCl stands for Sodium Chloride. Salt has many uses. For e.g. salt is used for cooking, salt is used during the production of toothpaste, plastic, soaps, bleaches, etc. The pH of salt is 7 which means that it is neutral in nature i.e. it is neither acidic nor basic.

Properties of Salt

- Salt is made up of Sodium and Chlorine.
- Salt has negatively charged ions (OH^-) and positively charged ions (H^+).
- Due to their oppositely charged ions, the ions are attracted towards each other with an electrostatic force of attraction which is called an ionic bond.
- An equal number of opposite charges makes the ionic compounds neutral with no charge.
- Salt water is a good conductor of electricity.
- Salts are ionic in nature due to the presence of ions.
- They are brittle, hard and crystalline solids.
- Salt is white, odorless and it has a salty taste.
- All potassium (K), ammonium (NH_4^+) and sodium (Na) salts are soluble in water (H_2O).
- Nitrites, nitrates, and bicarbonates can be dissolved in water.
- All metallic oxides, metallic carbonates, hydroxides, phosphates, and sulphides are insoluble in water.

Types of Salts

The following are the different kinds of salt:

- Normal salt:

Normal salts are electrically neutral. They are formed when acids and bases neutralize, and these salts don't have replaceable hydrogen or hydroxyl in their formula. Metallic ions replace the hydrogen ions completely. For e.g. NaCl, KNO_3, $CuSO_4$ etc.

- Basic salt:

Salt formed due to partial replacement of hydroxy radicals of a diacidic base or a triacidic base with an acid radical. This kind of acid consists of hydroxyl, metallic cation and anion of an acid. For e.g. basic Zinc chloride, basic magnesium chloride, ZnOHCL etc.

- Acidic salt:

If a polybasic acid is neutralized partly by a base, the salt formed is acidic. In other words, such salt is produced by the replacement of only a part of the acidic hydrogen of the polybasic acid by a metal. For e.g. $NaHSO_4$, NaHS, $NaHCO_3$ etc.

- Double salt:

Double salts have more than one cation or anion. They're a crystalline salt having the composition of a mixture of two simple salts but with a different crystal structure from either. For e.g. bromlite, potassium sodium tartrate, aluminium sulfacetate etc.

- Mixed salt:

A mixed salt is a salt made from more than one base or acid. For e.g. sodium potassium sulphate, bleaching powder etc.

- Complex salt:

Salt formed due to combining a saturated solution of simple salts followed by crystallization of the solution similar to double salts. For e.g. Sodium silver cyanide, potassium mercuric iodide etc.

Oxides

Oxides are chemical compounds with one or more oxygen atoms combined with another element (e.g. Li_2O). Oxides are binary compounds of oxygen with another element, e.g., CO_2, SO_2, CaO, CO, ZnO, BaO_2, H_2O, etc. These are termed as oxides because here, oxygen is in combination with only one element. Based on their acid-base characteristics oxides are classified as acidic, basic, amphoteric or neutral:

- An oxide that combines with water to give an acid is termed as an acidic oxide.
- The oxide that gives a base in water is known as a basic oxide.
- An amphoteric solution is a substance that can chemically react as either acid or base.
- However, it is also possible for an oxide to be neither acidic nor basic, but is a neutral oxide.

There are different properties which help distinguish between the three types of oxides. The term anhydride ("without water") refers to compounds that assimilate H_2O to form either an acid or a base upon the addition of water.

Acidic Oxides

Acidic oxides are the oxides of non-metals (Groups 14-17) and these acid anhydrides form acids with water.

- Sulfurous Acid:

$$SO_2 + H_2O \rightarrow H_2SO_3$$

- Sulfuric Acid:

$$SO_3 + H_2O \rightarrow H_2SO_4$$

- Carbonic Acid:

$$CO_2 + H_2O \rightarrow H_2CO_3$$

Acidic oxides are known as acid anhydrides (e.g., sulfur dioxide is sulfurous anhydride and sulfur trioxide is sulfuric anhydride) and when combined with bases, they produce salts, e.g.,

$$SO_2 + 2NaOH \rightarrow Na_2SO_3 + H_2O$$

Basic Oxides

Generally Group 1 and Group 2 elements form bases called base anhydrides or basic oxides e.g.,

$$K_2O(s) + H_2O(l) \rightarrow 2KOH(aq)$$

Basic oxides are the oxides of metals. If soluble in water, they react with water to produce hydroxides (alkalies) e.g.,

$$CaO + H_2O \rightarrow Ca(OH)_2$$
$$MgO + H_2O \rightarrow Mg(OH)_2$$
$$Na_2O + H_2O \rightarrow 2NaOH$$

These metallic oxides are known as basic anhydrides. They react with acids to produce salts, e.g.,

$$MgO + 2HCl \rightarrow MgCl_2 + H_2O$$
$$Na_2O + H_2O \rightarrow 2NaOH$$

Amphoteric Oxides

An amphoteric solution is a substance that can chemically react as either acid or base. For example, when HSO_4- reacts with water it will make both hydroxide and hydronium ions:

$$HSO_4^- + H_2O \rightarrow SO_4^{2-} + H_3O^+$$
$$HSO_4^- + H_2O \rightarrow H_2SO_4 + OH^-$$

Amphoteric oxides exhibit both basic as well as acidic properties. When they react with an acid, they produce salt and water, showing basic properties. While reacting with alkalies they form salt and water showing acidic properties.

- For example Z_nO exhibits basic behavior with HCl:

$$ZnO + 2HCl \rightarrow \underset{zinc\,chloride}{ZnCl_2} + H_2O(basic\,nature)$$

and acidic behavior with NaOH,

$$ZnO + 2NaOH \rightarrow \underset{sodium\,zincate}{Na_2ZnO_2} + H_2O(acidic\,nature)$$

- Similarly, Al_2O_3 exhibits basic behavior with H_2SO_4:

  $$Al_2O_3 + 3H_2SO_4 \rightarrow Al_2(SO_4)_3 + 3H_2O(basic\,nature)$$

  and acidic behavior with $NaOH$,

  $$Al_2O_3 + 2NaOH \rightarrow 2NaAlO_2 + H_2O(acidic\,nature) \cdot$$

Neutral Oxides

Neutral oxides show neither basic nor acidic properties and hence do not form salts when reacted with acids or bases, e.g., carbon monoxide (CO); nitrous oxide (N_2O); nitric oxide (NO), etc., are neutral oxides.

Peroxides and Dioxides

- Oxides: Group 1 metals react rapidly with oxygen to produce several different ionic oxides, usually in the form of M_2O. With the oyxgen exhibiting an oxidation number of -2.

  $$4Li + O_2 \rightarrow 2Li_2O$$

- Peroxides: Often Lithium and Sodium reacts with excess oxygen to produce the peroxide, M_2O_2. With the oxidation number of the oxygen equal to -1.

  $$H_2 + O_2 \rightarrow H_2O_2$$

- Superoxides: Often Potassium, Rubidium, and Cesium react with excess oxygen to produce the superoxide, MO_2. With the oxidation number of the oxygen equal to -1/2.

  $$Cs + O_2 \rightarrow CsO_2$$

A peroxide is a metallic oxide which gives hydrogen peroxide by the action of dilute acids. They contain more oxygen than the corresponding basic oxide, e.g., sodium, calcium and barium peroxides.

$$BaO_2 + H_2SO_4 \rightarrow BaSO_4 + H_2O_2$$
$$Na_2O_2 + H_2SO_4 \rightarrow Na_2SO_4 + H_2O_2$$

Dioxides like PbO_2 and MnO_2 also contain higher percentage of oxygen like peroxides and have similar molecular formulae. These oxides, however, do not give hydrogen peroxide by action with dilute acids. Dioxides on reaction with concentrated HCl yield Cl_2 and on reacting with concentrated H_2SO_4 yield O_2.

$$PbO_2 + 4HCl \rightarrow PbCl2 + Cl_2 + 2H_2O$$
$$2PbO_2 + 2H_2SO_4 \rightarrow 2PbSO_4 + 2H_2O + O_2$$

Compound Oxides

Compound oxides are metallic oxides that behave as if they are made up of two oxides, one that has a lower oxidation and one with a higher oxidation of the same metal, e.g.,

Red lead : $Pb_3O_4 = PbO_2 + 2PbO$

Ferro-ferric oxide : $Fe_3O_4 = Fe_2O_3 + FeO$

On treatment with an acid, compound oxides give a mixture of salts:

$$\underset{\text{Ferro-ferric oxide}}{Fe_3O_4} + 8HCl \rightarrow \underset{\text{ferric chloride}}{2FeCl_3} + \underset{\text{ferrous chloride}}{FeCl_2} + 4H_2O$$

Preparation of Oxides

Oxides can be generated via multiple reactions.

1. By direct heating of an element with oxygen: Many metals and non-metals burn rapidly when heated in oxygen or air, producing their oxides, e.g.:

$$2Mg + O_2 \xrightarrow{Heat} 2MgO$$
$$2Ca + O_2 \xrightarrow{Heat} 2CaO$$
$$S + O_2 \xrightarrow{Heat} SO_2$$
$$P_4 + 5O_2 \xrightarrow{Heat} 2P_2O_5$$

By reaction of oxygen with compounds at higher temperatures: At higher temperatures, oxygen also reacts with many compounds forming oxides, e.g.,

- Sulfides are usually oxidized when heated with oxygen:

$$2PbS + 3O_2 \xrightarrow{\Delta} 2PbO + 2SO_2$$
$$2ZnS + 3O_2 \xrightarrow{\Delta} 2ZnO + 2SO_2$$

- When heated with oxygen, compounds containing carbon and hydrogen are oxidized:

$$C_2H_5OH + 3O_2 \rightarrow 2CO_2 + 3H_2O$$

- By thermal decomposition of certain compounds like hydroxides, carbonates, and nitrates:

$$CaCO_3 \xrightarrow{\Delta} CaO + CO_2$$
$$2Cu(NO_3)_2 \xrightarrow{\Delta} 2CuO + 4NO_2 + O_2$$
$$Cu(OH)_2 \xrightarrow{\Delta} CuO + H_2O$$

2. By oxidation of some metals with nitric acid:

$$2Cu + 8HNO_3 \xrightarrow{Heat} 2CuO + 8NO_2 + 4H_2O + O_2$$

$$Sn + 4HNO_3 \xrightarrow{Heat} SnO_2 + 4NO_2 + 2H_2O$$

3. By oxidation of some non-metals with nitric acid:

$$C + 4HNO_3 \rightarrow CO_2 + 4NO_2 + 2H_2O$$

Trends in Acid-Base Behavior

The oxides of elements in a period become progressively more acidic as one goes from left to right in a period of the periodic table. For example, in third period, the behavior of oxides changes as follows:

$$\underbrace{Na_2O, MgO}_{Basic}\ \underbrace{Al_2O_3, SiO_2}_{Amphoteric}\ \underbrace{P_4O_{10}, SO_3, Cl_2O_7}_{Acidic}$$

If we take a closer look at a specific period, we may better understand the acid-base properties of oxides. It may also help to examine the physical properties of oxides, but it is not necessary. Metal oxides on the left side of the periodic table produce basic solutions in water (e.g. Na_2O and MgO). Non-metal oxides on the right side of the periodic table produce acidic solutions (e.g. Cl_2O, SO_2, P_4O_{10}). There is a trend within acid-base behavior: basic oxides are present on the left side of the period and acidic oxides are found on the right side.

Aluminum oxide shows acid and basic properties of an oxide, it is amphoteric. Thus Al_2O_3 entails the marking point at which a change over from a basic oxide to acidic oxide occurs. It is important to remember that the trend only applies for oxides in their highest oxidation states. The individual element must be in its highest possible oxidation state because the trend does not follow if all oxidation states are included. Notice how the amphoteric oxides (shown in blue) of each period signify the change from basic to acidic oxides.

Groups						
1	2	3	14	15	16	17
Li	Be	B	C	N	O	F
Na	Mg	Al	Si	P	S	Cl
K	Ca	Ga	Ge	As	Se	Br
Rb	Sr	In	Sn	Sb	Te	I
Cs	Ba	Tl	Pb	Bi	Po	At

The figure above show oxides of the s- and p-block elements. Purple: basic oxides; Blue: amphoteric oxides; Pink: acidic oxides.

Metal Hydrides

Metal hydrides are metals which have been bonded to hydrogen to form a new compound. Any hydrogen compound that is bonded to another metal element can effectively be called a metal hydride. Generally, the bond is covalent in nature, but some hydrides are formed from ionic bonds. The hydrogen has an oxidation number of -1. The metal absorbs the gas, which forms the hydride.

Examples of Metal Hydrides

The most common examples of metal hydrides include aluminum, boron, lithium borohydride and various salts. For example, aluminum hydrides include sodium aluminum hydride. There are a number of types of hydrides. This includes aluminum, beryllium, cadmium, caesium, calcium, copper, iron, lithium, magnesium, nickel, palladium, plutonium, potassium rubidium, sodium, thallium, titanium, uranium and zinc hydrides.

There are also many more complex metal hydrides suitable for various uses. These complex metal hydrides are often soluble in etherial solvents.

Metal Hydrides Classes

There are four classes of metal hydrides. The most common hydride is those that form with hydrogen, dubbed binary metal hydrides. There are only two compounds—hydrogen and metal. These hydrides are generally insoluble, being conductive.

Other types of metal hydrides are less common or known, including ternary metal hydrides, coordination complexes, and cluster hydrides.

Hydride Formulation

Metal hydrides are formed via one of four syntheses. The first is the hydride transfer, which is metathesis reactions. Then there are elimination reactions, which includes the elimination of beta-hydride and alpha-hydride.

The third is oxidative additions, which is generally the transition of dihydrogen to a low valent metal center. The fourth is heterolytic cleavage of dihydrogen, this happens when hydrides are formed when metal complexes are treated with hydrogen in the presence of a base.

There are a variety of complexes, including Mg-based hayrides, known for their storage capacity and being thermally stable. The testing of such compounds under high pressure has opened it up hydrides to new uses. The high pressure prevents thermal decomposition.

In terms of bridging hydrides, metal hydrides with terminal hydrides are normal, with most being oligomeric. The classical thermal hydride involves binding metal and hydrogen. Meanwhile, bridging ligand is classical bridging that uses hydrogen to bind two metals. Then there is dihydrogen complex bridging that's non-classical. This happens when bi-hydrogen bonds with a metal.

The number of hydrogen must match the oxidation number of the metal. For example, the symbol for calcium hydride is CaH_2, but for Tin it's SnH_4.

Uses for Metal Hydrides

Metal hydrides are often used in fuel cell applications that use hydrogen as a fuel. Nickel hydrides are often found in various types of batteries, particularly NiMH batteries. Nickel metal hydride batteries rely on hydrides of rare-earth intermetallic compounds, such as lanthanum or neodymium bonded with cobalt or manganese. Lithium hydrides and sodium borohydride both serve as reducing agents in chemistry applications. Most hydrides behave as reducing agents in chemical reactions.

Beyond fuel cells, metal hydrides are used for their hydrogen storage and compressors capabilities. Metal hydrides are also used for heat storage, heat pumps, and isotope separation. The uses include sensors, activators, purification, heat pumps, thermal storage, and refrigeration.

Chalcogen and Chalcogenides

Sulfur, selenium, and tellurium are called chalcogens. Simple substances and compounds of oxygen and of the elements of this group in the later periods have considerably different properties. As a result of having much smaller electronegativities than oxygen, they show decreased ionicity and increased bond covalency, resulting in a smaller degree of hydrogen bonding. Because they have available *d* orbitals, chalcogens have increased flexibility of valence and can easily bond to more than two other atoms.

Catenation is the bonding between the same chalcogen atoms, and both simple substances and ions of chalcogens take a variety of structures.

The major isotopes of sulfur are ^{32}S (95.02% abundance), ^{33}S (0.75%), ^{34}S (4.21%), and ^{36}S (0.02%), and there are also six radioactive isotopes. Among these, ^{33}S ($I = 3/2$) can be used for NMR. Since the isotope ratio of sulfurs from different locations differs, the accuracy of the atomic weight is limited to 32.07+0.01. Because the electronegativity of sulfur ($\chi = 2.58$) is much smaller than that of oxygen ($\chi = 3.44$) and sulfur is a soft element, the iconicity in the bonds of sulfur compounds is low and hydrogen bonding is not important. Elemental sulfur has many allotropes, such as S_2, S_3, S_6, S_7, S_8, S_9, S_{10}, S_{11}, S_{12}, S_{18}, S_{20}, and S_∞, reflecting the catenation ability of sulfur atoms.

Elemental sulfur is usually a yellow solid with a melting point of 112.8 °C called orthorhombic sulfur (a sulfur). Phase-transition of this polymorph produces monoclinic sulfur (b sulfur) at 95.5 °C. It was established in 1935 that these are crown-like cyclic molecules. Being molecular, they dissolve well in organic solvents, such as CS_2. Not only 8-membered rings but also $S_{6\text{-}20}$ rings are known, and the helix polymer of sulfur is infinitely annular sulfur. Diatomic molecular S_2 and triatomic molecular S_3 exist in the gaseous phase. When sulfur is heated, it liquifies and becomes a rubber-like macromolecule on cooling. The diversity of structures of catenated sulfur is also seen in the structures of the polysulfur cations or anions resulting from the redox reactions of the catenated species.

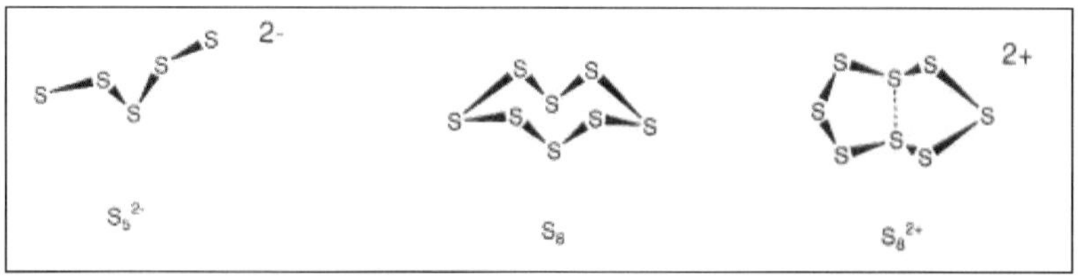

Structures of S_5^{2-}, S_8, and S_8^{2+}.

Selenium is believed to have six isotopes. ^{80}Se (49.7%) is the most abundant and ^{77}Se, with nuclear spin $I = 1/2$ is useful in NMR. The accuracy of atomic weight of selenium, 78.96+0.03, is limited to two decimal places because of composition change of its isotopes. Among many allotropes of selenium, so-called red selenium is an Se_8 molecule with a crown-like structure and is soluble in CS_2. Gray metallic selenium is a polymer with a helical structure. Black selenium, which is a complicated polymer, is also abundant.

Tellurium also has eight stable isotopes and an atomic weight of 127.60 + 0.03. ^{130}Te (33.8%) and ^{128}Te (31.7%) are the most abundant isotopes, and ^{125}Te and ^{123}Te with $I = 1/2$ can be used in NMR. There is only one crystalline form of tellurium, which is a spiral chain polymer that shows electric conductivity.

Polyatomic Chalcogen Cations and Anions

Although it has long been recognized that solutions of chalcogen elements in sulfuric acid showed beautiful blue, red, and yellow colors, the polycationic species that give rise to these colors, S_4^{2+}, S_6^{2+}, S_6^{4+}, S_8^{2+}, S_{10}^{2+}, S_{19}^{2+}, or those of other chalcogen atoms, have been isolated by the reaction with AsF_5, *etc.* and their structures determined. For example, unlike neutral S_8, S_8^{2+} takes a cyclic structure that has a weak coupling interaction between two transannular sulfur atoms.

On the other hand, alkali metal salts Na_2S_2, K_2S_5, and alkaline earth metal salt BaS_3, a transition-metal salt $[Mo_2(S_2)_6]^{2-}$, a complex $Cp_2W(S_4)$, *etc.* of polysulfide anions S_x^{2-} (x = 1-6), in which the sulfur atoms are bonded mutually have been synthesized and their structures determined. As is evident from the fact that elemental sulfur itself forms S_8 molecules, sulfur, unlike oxygen, tends to catenate. Therefore, formation of polysulfide ions, in which many sulfur atoms are bonded, is feasible, and a series of polysulfanes H_2S_x (x = 2-8) has actually been synthesized.

Metal Sulfides

Stratified disulfides, MS_2, are important in transition metal sulfides. They show two types of structures. One has a metal in a triangular prismatic coordination environment, and the other has a metal in an octahedral coordination environment.

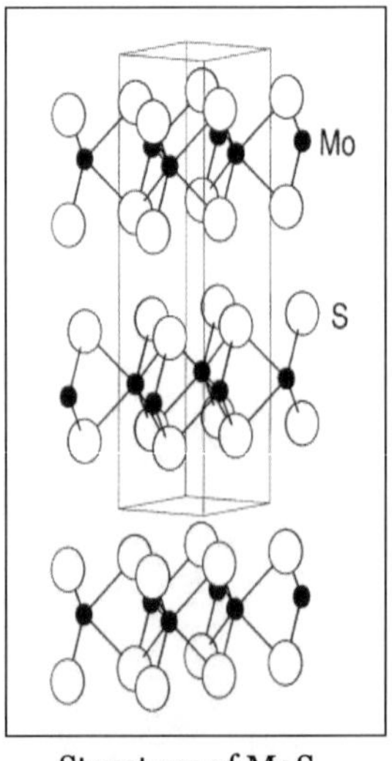

Structure of MoS_2.

MoS_2 is the most stable black compound among the molybdenum sulfides. L. Pauling determined the structure of MoS_2 in 1923. The structure is constructed by laminating two sulfur layers between which a molybdenum layer is intercalated. Alternatively, two sulfur layers are stacked and a molybdenum layer is inserted between them. Therefore, the coordination environment of each molybdenum is a triangular prism of sulfur atoms. Since there is no bonding interaction between sulfur layers, they can easily slide, resulting in graphite-like lubricity. MoS_2 is used as a solid lubricant added to gasoline, and also as a catalyst for hydrogenation reactions.

ZrS_2, TaS_2, *etc.* take the CdI_2 type structure containing metal atoms in an octahedral coordination environment constructed by sulfur atoms.

Chevrel phase compounds: There are superconducting compounds called Chevrel phases which are important examples of the chalcogenide compounds of molybdenum. The general formula is described by $M_xMo_6X_8$ (M = Pb, Sn, and Cu; X = S, Se, and Te), and six molybdenum atoms form a regular octahedral cluster, and eight chalcogenide atoms cap the eight triangular faces of the cluster. The cluster units are connected 3-dimensionally. Since the cluster structure of molybdenum atoms is similar to that of molybdenum dichloride, $MoCl_2$, (= $(Mo_6Cl_8)Cl_2Cl_{4/2}$), the structural chemistry of these compounds has attracted as much attention as their physical properties.

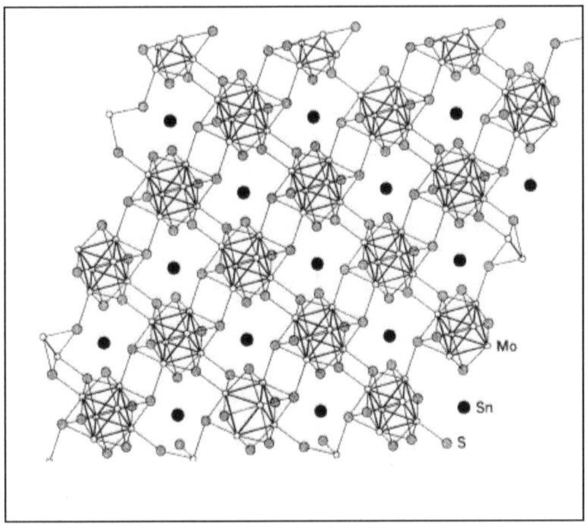

Structure of $SnMno_6S_8$.

Halogens and Halides

The origin of halogen is the Greek word meaning the production of salt by direct reaction with a metal. Since their reactivity is very high, halogens are found in nature only as compounds. The basic properties of halogens are shown in tables. The electron configuration of each halogen atom is ns^2np^5, and they lack one electron from the closed-shell structure of a rare gas. Thus a halogen atom emits energy when it gains an electron. Namely, the enthalpy change of the reaction is negative. Although electron affinity is defined as the energy change of gaining an electron, a positive sign is customarily used. In order to be consistent with the enthalpy change, a negative sign would be appropriate.

$$X(g)+e^- \rightarrow X^-(g)$$

Table - Properties of halogens			
	Ionization energy (kJ mol-1)	Electronegativity χP	Ionic radius r(X-) (pm)
F	1680.6	3.98	133
Cl	1255.7	3.16	181
Br	1142.7	2.996	196
I	1008.7	2.66	220

Table - Properties of halogen molecules				
	Interatomic distance r(X-X) (pm)	mp °C	bp °C	Color
F_2	143	-218.6	-188.1	Colorless gas
Cl_2	199	-101.0	-34.0	Yellow green gas
Br_2	228	-7.75	59.5	Dark red liquid
I_2	266	113.6	185.2	Dark violet solid

The electron affinity of chlorine (348.5 kJ mol^{-1}) is the largest and fluorine (332.6 kJ mol^{-1}) comes between chlorine and bromine (324.7 kJ mol^{-1}). The electronegativity of fluorine is the highest of all the halogens.

Since halogens are produced as metal salts, simple substances are manufactured by electrolysis. Fluorine only takes the oxidation number -1 in its compounds, although the oxidation number of other halogens can range from -1 to +7. Astatine, at, has no stable nuclide and little is known about its chemical properties.

Manufacture of Halogen

Fluorine has the highest reduction potential (E = +2.87 V) and the strongest oxidizing power among the halogen molecules. It is also the most reactive nonmetallic element. Since water is oxidized by F_2 at much lower electrode potential (+1.23 V), fluorine gas cannot be manufactured by the electrolysis of aqueous solutions of fluorine compounds. Therefore, it was a long time before elemental fluorine was isolated, and F. F. H. Moisson finally succeeded in isolating it by the electrolysis of KF in liquid HF. Fluorine is still manufactured by this reaction.

Chlorine, which is especially important in inorganic industrial chemistry, is manufactured together with sodium hydroxide. The basic reaction for the production of chlorine is electrolysis of an aqueous solution of NaCl using an ion exchange process. In this process, chlorine gas is generated in an anodic cell containing brine and Na^+ moves through an ion exchange membrane to the cathodic cell where it pairs with OH^- to become an aqueous solution of NaOH.

Bromine is obtained by the oxidation of Br- with chlorine gas in saline water. Iodine is similarly produced by passing chlorine gas through saline water containing I^- ions. Since natural gas is found in Japan together with underground saline water containing I^-. Japan is one of the main countries producing iodine.

1. Anomalies of fluorine:

Molecular fluorine compounds have very low boiling points. This is due to the difficulty of polarization as a result of the electrons being strongly drawn to the nuclei of fluorine atoms. Since the electronegativity of fluorine is highest ($\chi = 3.98$) and electrons shift to F, resulting in the high acidity of atoms bonded to F. Because of the small ionic radius of F^-, high oxidation states are stabilized, and hence low oxidation compounds like CuF are unknown, in contrast with the compounds such as IF_7 and PtF_6.

2. Pseudohalogens:

Since the cyanide ion CN^-, the azide ion N^{3-}, and the thiocyanate ion SCN^-, etc. form compounds similar to those of halide ions, they are called pseudohalide ions. They form psudohalogen molecules such as cyanogene $(CN)_2$, hydrogen cyanide HCN, sodium thiocyanate NaSCN, etc. Fine-tuning electronic and steric effects that are impossible with only halide ions make pseudohalogens useful also in transition metal complex chemistry.

3. Polyhalogens:

Besides the usual halogen molecules, mixed halogen and polyhalogen molecules such as BrCl, IBr, ICl, ClF_3, BrF_5, IF_7 etc. also exist. Polyhalogen anions and cations such as I_3^-, I_5^-, I_3^+, and I_5^+, are also known.

Oxygen Compounds

Although many binary oxides of halogens (consisting only of halogen and oxygen) are known, most are unstable. Oxygen difluoride OF_2 is the most stable such compound. This is a very powerful fluorinating agent and can generate plutonium hexafluoride PuF_6 from plutonium metal. While oxygen chloride, Cl_2O, is used for bleaching pulp and water treatment, it is generated in situ from ClO_3^-, since it is unstable.

Hypochlorous acid, HClO, chlorous acid, $HClO_2$, chloric acid, $HClO_3$, and perchloric acid, $HClO_4$ are oxoacids of chlorine and especially perchloric acid is a strong oxidizing agent as well as being a strong acid. Although analogous acids and ions of other halogens had been known for many years, BrO_4^- was synthesized as late as 1968. Once it was prepared it turned out to be no less stable than ClO_4^- or IO_4^-, causing some to wonder why it had not been synthesized before. Although ClO_4^- is often used for crystallizing transition metal complexes, it is explosive and should be handled very carefully.

Halides of Nonmetals

Halides of almost all nonmetals are known, including fluorides of even the inert gases krypton, Kr, and xenon, Xe. Although fluorides are interesting for their own unique characters, halides are generally very important as starting compounds for various compounds of nonmetals by replacing halogens in inorganic syntheses table.

Table: Typical chlorides and fluorides of main group elements.

	1	2	12	13	14	15	16	17	18
2	LiCl	$BeCl_2$	BF_3	CCl_4	NF_3	OF_2			
3	NaCl	$MgCl_2$		$AlCl_3$	$SiCl_4$	PCl_3PCl_5	$S_2Cl_2SF_6$	ClF_3ClF_5	
4	KCl	$CaCl_2$	$ZnCl_2$	$GaCl_3$	GeF_2GeCl_4	$AsCl_3AsF_5$	$Se_2Cl_2SeF_5$	BrF_3BrF_5	KrF_2
5	RbCl	$SrCl_2$	$CdCl_2$	$InClInCl_3$	$SnCl_2SnCl_4$	$SbCl_3SbF_5$	$Te_4Cl_{16}TeF_6$	IF_5IF_7	XeF_2XeF_6
6	CsCl	$BaCl_2$	$Hg_2Cl_2HgCl_2$	$TlClTlCl_3$	$PbCl_2PbCl_4$	$BiCl_3BiF_5$			

Boron trifluoride, BF_3, is a colorless gas (mp -127 °C and bp -100 °C) that has an irritating odor and is poisonous. It is widely used as an industrial catalyst for Friedel-Crafts type reactions. It is also used as a catalyst for cationic polymerization. It exists in the gaseous phase as a triangular monomeric molecule, and forms Lewis base adducts with ammonia, amines, ethers, phosphines, etc. because of its strong Lewis acidity. Diethylether adduct, $(C_2H_5)_2O:BF_3$, is a distillable liquid and is used as a common reagent. It is a starting compound for the preparation of diborane, B_2H_6. Tetrafluoroborate, BF_4^-, is a tetrahedral anion formed as a adduct of BF_3 with a base F^-. Alkali metal salts, a silver salt and $NOBF_4$ as well as the free acid HBF_4 contain this anion. Since its coordination ability is very weak, it is used in the crystallization of cationic complexes of transition metals as a counter anion like ClO_4^-. $AgBF_4$ and $NOBF_4$ are also useful for 1-electron oxidation of complexes.

Tetrachlorosilane, $SiCl_4$, is a colorless liquid (mp -70 °C and bp 57.6 °C). It is a regular tetrahedral molecule, and reacts violently with water forming silicic acid and hydrochloric acid. It is useful as a raw material for the production of pure silicon, organic silicon compounds, and silicones.

Phosphorus trifluoride, PF_3, is a colorless, odorless, and deadly poisonous gas (mp -151.5 °C and bp -101.8 °C). This is a triangular pyramidal molecule. Because it is as electron-attracting as CO, it acts as a ligand forming metal complexes analogous to metal carbonyls.

Phosphorus pentafluoride, PF_5, is a colorless gas (mp -93.7 °C and bp -84.5 °C). It is a triangular bipyramidal molecule and should have two distinct kinds of fluorine atoms. These fluorines exchange positions so rapidly that they are indistinguishable by ^{19}F NMR. It was the first compound with which the famous Berry's pseudorotation was discovered as an exchange mechanism for axial and equatorial fluorine atoms. The hexafluorophosphate ion, PF_6^-, as well as BF_4^- is often used as a counter anion for cationic transition metal complexes. $LiPF_6$ and R_4NPF_6 can be used as supporting electrolytes for electrochemical measurements.

Phosphorus trichloride, PCl_3, is a colorless fuming liquid (mp -112 °C and bp 75.5 °C). It is a triangular pyramidal molecule and hydrolyzes violently. It is a soluble in organic solvents. It is used in large quantities as a raw material for the production of organic phosphorus compounds.

Phosphorus pentachloride, PCl_5, is a colorless crystalline substance (sublimes but decomposes at 160 °C) It is a triangular bipyramidal molecule in the gaseous phase, but it exists as an ionic crystal $[PCl_4]^+ [PCl_6]^-$ in the solid phase. Although it reacts violently with water and becomes phosphoric acid and hydrochloric acid, it dissolves in carbon disulfide and carbon tetrachloride. It is useful for cchlorination of organic compounds.

Arsenic pentafluoride, AsF_5, is a colorless gas (mp -79.8 °C and bp -52.9 °C). It is a triangular bipyramidal molecule. Although it hydrolyzes, it is soluble in organic solvents. As it is a strong electron acceptor, it can form electron donor-acceptor complexes with electron donors.

Sulfur hexafluoride, SF_6, is a colorless and odorless gas (mp -50.8 °C and sublimation point -63.8 °C) It is a hexacoordinate octahedral molecule. It is chemically very stable and hardly soluble in water. Because of its excellent heat-resisting property, incombustibility, and corrosion resistance, it is used as a high voltage insulator.

Sulfur chloride, S_2Cl_2, is an orange liquid (mp -80 °C and bp 138 °C). It has a similar structure to hydrogen peroxide. It is readily soluble in organic solvents. It is important as an industrial inorganic compound, and is used in large quantities for the vulcanization of rubber etc.

Metal Halides

Many metal halides are made by the combination of about 80 metallic elements and four halogens. Since there are more than one oxidation state especially in transition metals, several kinds of halides are known for each transition metal. These halides are most important as starting materials of the preparation of metal compounds, and the inorganic chemistry of metal compounds depends on metal halides. There are molecular, 1-dimensional chain, 2-dimensional layer, and 3-dimensional halides but few of them are molecular in crystalline states. It should be noted that the anhydrous transition metal halides are usually solid compounds and hydrates are coordination compounds with water ligands. As the dimensionality of structures is one of the most interesting facets of structural or synthetic chemistry, typical halides are described in order of their dimensionality.

Table: Typical chlorides and fluorides of transition metals.

Oxidation Number	3	4	5	6	7	8	9	10	11
+1	ScCl YCl LaCl	ZrCl HfCl							CuCl AgCl AuCl
+2		$TiCl_2$	VCl_2	$CrCl_2$ $MoCl_2$ WCl_2	$MnCl_2$	$FeCl_2$ $RuCl_2$	$CoCl_2$	$NiCl_2$ $PdCl_2$ $PdCl_2$	$CuCl_2$
+3	ScF_3 YCl_3 LaF_3	$TiCl_3$ $ZrCl_3$	VCl_3	$CrCl_3$ $MoCl_3$ WCl_3	$ReCl_3$	$FeCl_3$ $RuCl_3$ $OsCl_3$	CoF_3 $RhCl_3$ $IrCl_3$		$AuCl_3$
+4		$TiCl_4$ $ZrCl_4$ $HfCl_4$	VCl_4 $NbCl_4$ $TaCl_4$	CrF_4 $MoCl_4$ WCl_4	$ReCl_4$			$PtCl_4$	
+5			VF_5 $NbCl_5$ $TaCl_5$	CrF_5 $MoCl_5$ WCl_5	$ReCl_5$	OsF_5	IrF_5	PtF_5	
+6					ReF_6	OsF_6	IrF_6	PtF_6	
+7					ReF_7	OsF_7			

Molecular Halides

Mercury(II) Chloride ($HgCl_2$)

It is a colorless crystal soluble in water and ethanol. It is a straight, three-atomic molecule in the free-state. However, in addition to two chlorine atoms bonded to mercury, four additional chlorine atoms of adjacent molecules occupy coordination sites and the mercury is almost hexacoordinate in the crystalline state. The compound is very toxic and used for preserving wood, etc.

Aluminum Trichloride ($AlCl_3$)

A colorless crystal (mp 190 °C (2.5 atm) and bp 183 °C) that sublimes when heated. It is soluble in ethanol and ether. It is a Lewis acid and forms adducts with various bases. It is a molecule consisting of the dimer of tetracoordinate aluminium with chlorine bridges in the liquid and gaseous phases figure below, and takes a lamellar structure when crystalline. It is used as a Lewis acid catalyst of Friedel-Crafts reactions, etc.

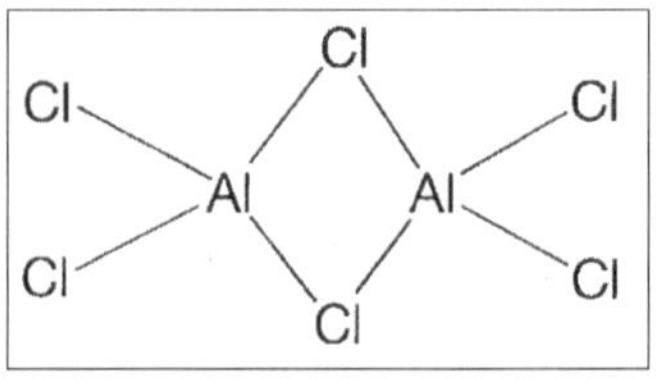

Structure of aluminum chloride.

Tin (IV) Chloride ($SnCl_4$)

A colorless liquid (mp -33 °C and bp 114 °C). In the gaseous state, it is a tetrahedral molecule.

Titanium (IV) Chloride ($TiCl_4$)

A colorless liquid (mp -25 °C and bp 136.4 °C). The gaseous molecule is a tetrahedron similar to tin(IV) chloride. It is used as a component of the Ziegler Natta catalyst.

Chain-like Halides

Gold (I) Iodide (AuI)

Yellow white solid. Two iodines coordinate to gold, and the compound has a zigzag 1-dimensional chain structure.

Beryllium Chloride ($BeCl_2$)

A colorless crystal (mp 405 °C and bp 520 °C). It is deliquescent and soluble in water and ethanol. The tetra-coordinated beryllium forms a 1-dimensional chain via chlorine bridges figure below. In the gaseous phase, it is a straight two-coordinate molecule. It is a Lewis acid and is used as a catalyst for Friedel-Crafts reactions.

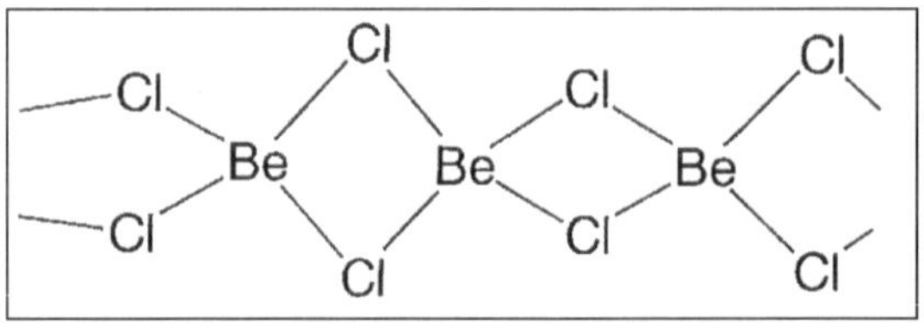

Structure of beryllium chloride.

Palladium Chloride ($PdCl_2$)

A dark red solid. In α type, the four-coordinate palladium forms a 1-dimensional chain with double bridges of chlorines. The dihydrate is deliquescent and soluble in water, ethanol, acetone, etc.

When it is dissolved in hydrochloric acid, it becomes four-coordinate square-planar $[PdCl_4]^{2-}$. It is used as the catalyst for the Wacker process, which is an olefin oxidation process, or in various catalysts for organic syntheses.

Zirconium Tetrachloride (IV) ($ZrCi_4$)

A colorless crystal (it sublimes above 331 °C). The zirconium is octahedrally coordinated and forms a zigzag chain via chlorine bridges figure below. It is hygroscopic and soluble in water, ethanol, etc. It is used as a Friedel-Crafts catalyst and as a component of olefin polymerization catalysts.

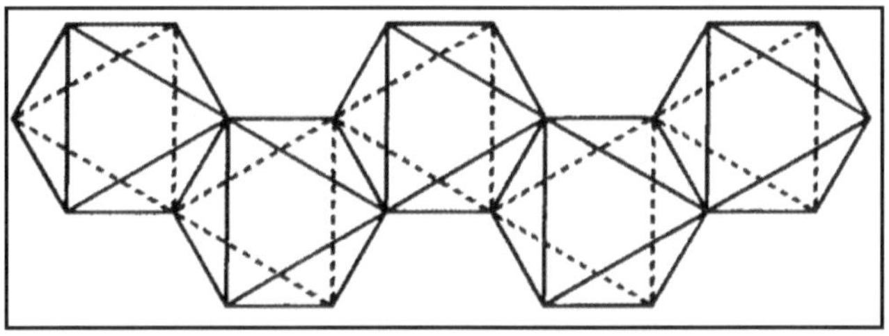

Structure of zirconium tetrachloride.

Stratified Halides

Cadmium Iodide (CdI_2)

A colorless crystal (mp 388 °C and bp 787 °C). It has a cadmium iodide structure where the layers of edge-shared CdI_6 octahedral units are stratified figure below. In the gaseous phase, it comprises straight three atomic molecules. It dissolves in water, ethanol, acetone, etc.

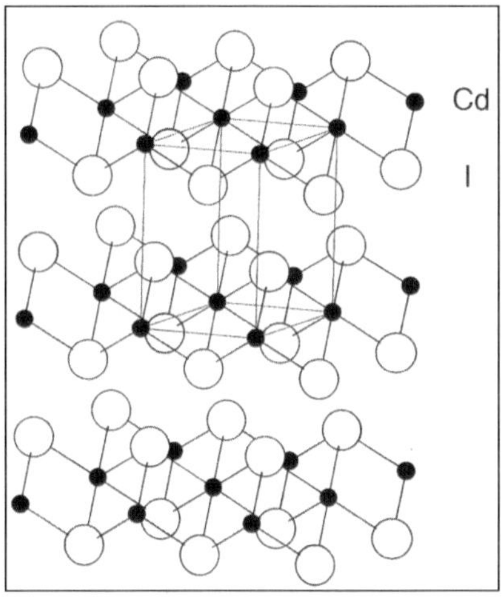

Layers of cadmium and iodine in the cadmium iodide structure.

Cobalt (II) Chloride ($CoCl_2$)

Blue crystals (mp 735 °C and bp 1049 °C). It has the cadmium chloride structure. It is hygroscopic and becomes light red when water is absorbed. It is soluble also in ethanol and acetone. The hexa-hydrate is red and is a coordination compound in which water molecules are ligands.

Iron (II) Chloride ($FeCl_2$)

Greenish yellow crystals (mp 670-674 °C). It has the cadmium chloride structure, and is soluble in water and ethanol. The hydrates, which are coordinated by various numbers (6, 4, 2) of water molecules, are precipitated from aqueous solutions of hydrochloric acid.

Iron (III) Chloride ($FeCl_3$)

Dark brown crystals (mp 306 °C and sublimes). It has a lamellar structure in which iron is octahedrally surrounded by six chlorine ligands. In the gaseous phase, it has a dimeric structure bridged by chlorine atoms similar to that of aluminum chloride.

3-dimensional Structure Halides

Sodium Chloride (NaCl)

A colorless crystal (mp 801 °C and bp 1413 °C). It is the original rock salt-type structure. In the gaseous phase, this is a two-atom molecule. Although it is soluble in glycerol as well as water, it hardly dissolves in ethanol. Large single crystals are used as prisms for infrared spectrometers.

Cesium Chloride (CsCl)

A colorless crystal (mp 645 °C, bp 1300 °C). Although it has the cesium chloride type structure, it changes to the rock salt structure at 445 °C. In the gaseous phase, it is a two-atom molecule.

Copper (I) Chloride (CuCl)

A colorless crystal (mp 430 °C and bp 1490 °C) It has the zinc blende structure and four chlorines tetrahedrally coordinate to copper.

Calcium Chloride ($CaCl_2$)

A colorless crystal (mp 772 °C and bp above 1600 °C). It has a deformed rutile-type structure and calcium is octahedrally surrounded by six chlorines. It is soluble in water, ethanol, and acetone. It is deliquescent and used as a desiccant. Hydrates in which 1, 2, 4, or 6 water molecules are coordinated are known.

Calcium Fluoride (CaF_2)

A colorless crystal (mp 1418 °C and bp 2500 °C). It has the fluorite type structure. It is the most important raw material for fluorine compounds. Good quality crystals are used also as spectrometer prisms and in photographic lenses.

Chromium (II) Chloride ($CrCl_2$)

A colorless crystal (mp 820 °C and sublimes). It has a deformed rutile-type structure. It dissolves well in water giving a blue solution.

Chromium (III) Chloride ($CrCl_3$)

Purplish red crystal (mp 1150 °C and decomposes at 1300 °C). Cr^{3+} occupies two thirds of the octahedral cavities in every other layer of Cl^- ions, which are hexagonally close-packed. It is insoluble in water, ethanol, and acetone.

Hydrides and Metal Hydrides

Metal-hydrogen bonds, also known (misleadingly) as metal hydrides, are ubiquitous X-type ligands in organometallic chemistry. There is much more than meets the eye to most M-H bonds: although they're simple to draw, they vary enormously in polarization and pKa. They may be acidic or hydridic or both, depending on the nature of the metal center and the reaction conditions. In this post, we'll develop some heuristics for predicting the behavior of M-H bonds and discuss their major modes of reactivity (acidity, radical reactions, migratory insertion, etc.)

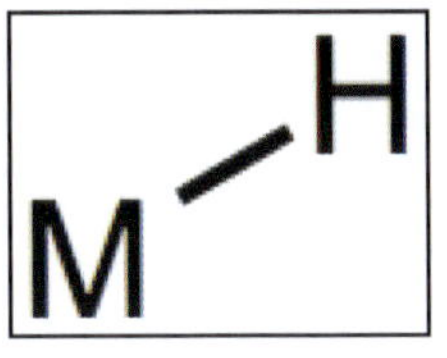

General Properties

Metal hydrides run the gamut from nucleophilic/basic to electrophilic/acidic. Then again, the same can be said of X–H bonds in organic chemistry, which may vary from mildly nucleophilic (consider Hantzsch esters and NADH) to extremely electrophilic (consider triflic acid). As hydrogen is what it is in both cases, it's clear that the nature of the X fragment—more specifically, the stability of the charged fragments X^+ and X^-–dictate the character of the X–H bond. Compare the four equilibria outlined below—the stabilities of the ions dictate the position of each equilibrium. By now we shouldn't find it surprising that the highly π-acidic $W(CO)_5$ fragment is good at stabilizing negative charge; for a similar reason, the $ZrCp_2Cl$ fragment can stabilize positive charge.

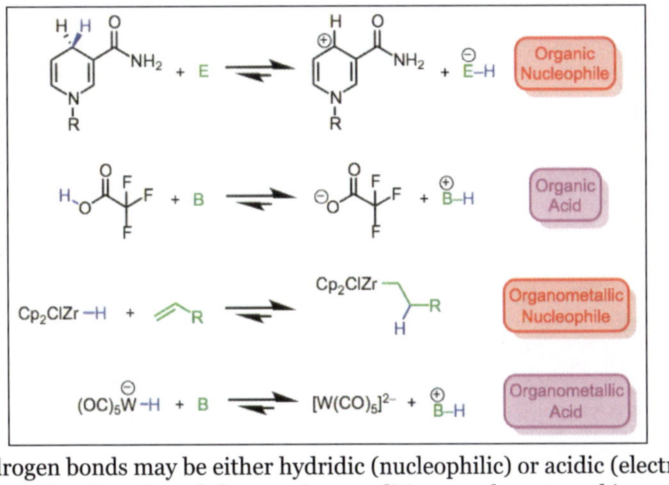

Metal-hydrogen bonds may be either hydridic (nucleophilic) or acidic (electrophilic). The nature of other ligands and the reaction conditions are keys to making predictions.

Let's turn our attention now to homolytic M–H bond strength. A convenient thermodynamic cycle allows us to use the acidity of M–H and the oxidation potential of its conjugate base in order to determine bond strength. This clever method, employed by Tilset and inspired by the inimitable Bordwell, uses the cycle in the figure below. BDE values for some complexes are provided as well. From the examples provided, we can see that bond strength increases down a group in the periodic table. This trend and the idea that bridging hydrides have larger BDEs than terminal M–H bonds are just about the only observable trends in M–H BDE.

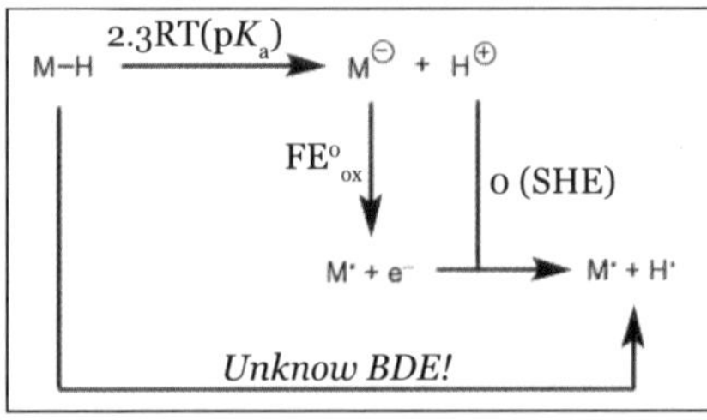

Complex	BDE (kcal/mol)
$CpCr(CO)_3H$	61
$CpMo(CO)_3H$	69
$CpW(CO)_3H$	72
$Mn(CO)_5H$	68
$Re(CO)_5H$	75

A clever cycle for determining BDEs from other known quantities, with select BDE values.

Why is knowing M-H BDEs useful? For one thing, the relative BDEs of M-C and M-H bonds determine the thermodynamics of β-hydride elimination, which results in the replacement of a covalent M-C bond with an M-H bond. Secondly, complexes containing weak M-H bonds are often good hydrogen transfer agents and may react with organic radicals and double bonds, channeling stannane and silane reductants from organic chemistry.

Hydricity refers to the tendency of a hydride ligand to depart as H^-. A similar thermodynamic cycle relates the energetics of losing H^- to the oxidation potentials of the conjugate base and the oxidized conjugate base; however, this method is complicated by the fact that hydride loss establishes an open coordination site. Hydricities are somewhat predictable from the electronic and steric properties of the metal center: Inclusion of electron-donating ligands tends to increase hydricity, while electron-withdrawing or acidic ligands tend to decrease it. For five-coordinate hydrides that form 16-electron, a square planar complex upon loss of hydride, the bite angle of chelating phosphines plays an interesting role. As bite angle increases, hydricity does as well.

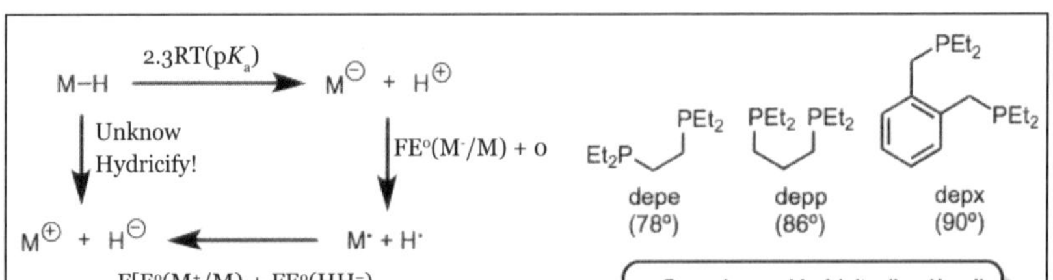

Complex	Hydricity (kcal/mol)
$HPd(depe)_2^+$	43
$HPd(depp)_2^+$	55
$HPd(depx)_2^+$	61

A thermodynamic cycle for hydricity, with some examples. Hydricity and bite angle are well correlated in five-coordinate palladium hydrides.

Bridging hydrides are an intriguing class of ligands. A question to ponder: how can a ligand associated with only two electrons possibly bridge two metal centers? How can two electrons hold three atoms together? Enter the magic of three-center, two-electron bonding. We can envision the M–H sigma bond as an electron donor itself. With this in mind, we can imagine that hydrides are able to bind end-on to one metal (like a standard X-type ligand) and side-on to another (like an L-type π system ligand, but using sigma electrons instead).

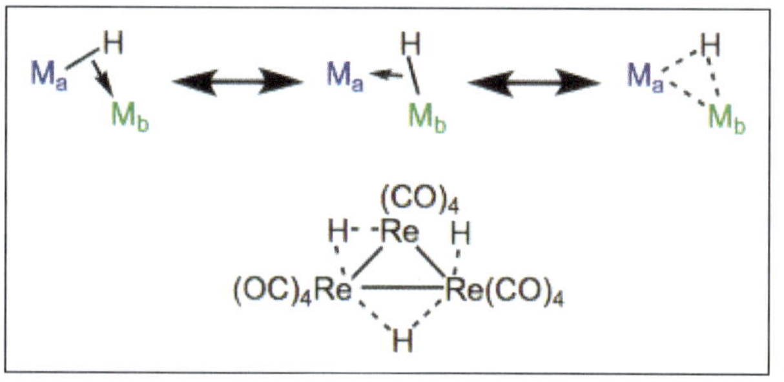

Resonance forms of bridging hydrides, with an example.
Sigma complexes like these show up in other contexts, too.

Consistent with the idea that bridging is the result of "end-on + side-on" bonding, bond angles of bridging hydrides are never 180°.

Synthesis

Here we'll discuss four ways to make hydrides: Metal protonation, oxidative addition of H_2, addition of nucleophilic main-group hydrides (borohydrides, aluminum hydrides, and silanes), and β-hydride elimination.

Just as in organic chemistry, the basicity of an organometallic complex is inversely related to the acidity of its conjugate base. Furthermore, charges have a predictable effect on the basicity of organometallic complexes: negatively charged complexes lacking π-acidic ligands are highly basic. Even neutral complexes containing strong donor ligands, like the tungsten complex below, can be protonated effectively. Notice that protonation is actually a kind of oxidative addition—the oxidation state of Fe in the first reaction below goes from -2 to 0 to +2. All coordination events of isolated electrophiles can also be viewed in this light. Reactions of this type are sometimes called oxidative ligations to distinguish them from oxidative addition reactions, which involve the addition of two ligands to the metal center with oxidation.

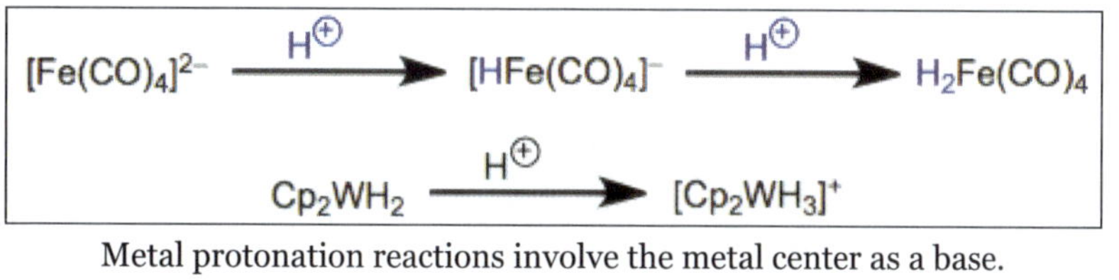

Metal protonation reactions involve the metal center as a base.

Contrast oxidative ligation with the oxidative addition of dihydrogen (H_2), a second method for the synthesis of hydride complexes. A key requirement here is that the starting metal center is at least d2—two electrons are formally lost from the metal center, and metals can't possess a negative number of d electrons. An open coordination site on the starting material must also be present (or possible through ligand dissociation). The reaction below is a standard example of the addition of H_2 to Vaska's complex, but there are some funky variations on this theme. These riffs include third-order homolytic cleavage of H_2 by two metal radicals, and oxidative addition followed by deprotonation by the starting complex (apparent heterolytic cleavage).

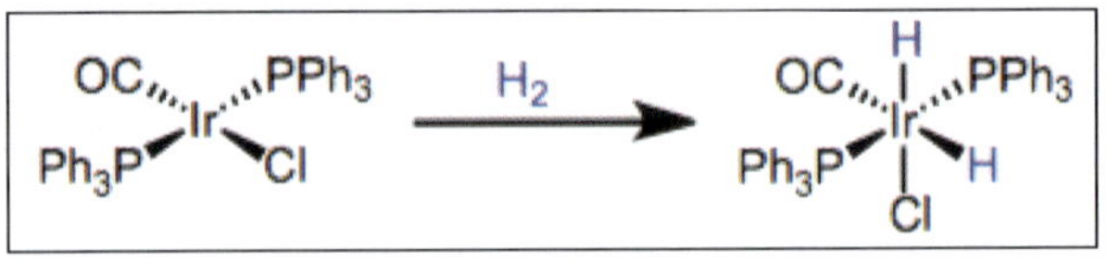

Oxidative addition of dihydrogen for the synthesis of metal hydrides.

Main-group hydrides like borohydrides and aluminum hydrides are great sources of H^- for organometallic complexes. These reactions seem more natural than metal protonations, since we often think of metals as electropositive or electrophilic species. Indeed, the combination of main-group sources of nucleophilic hydride with complexes containing metal–leaving group bonds is a very general method for the synthesis of metal hydride complexes.

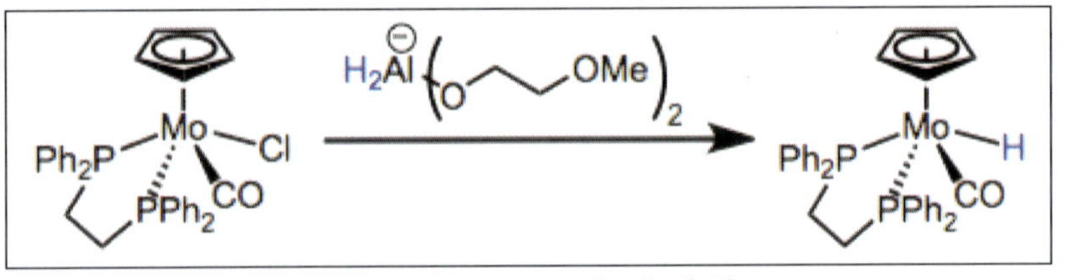

Nucleophilic substitution with a hydride source.

β-hydride elimination forms metal hydride complexes and double bonds within organic ligands. Alkoxide ligands are commonly used for this purpose—elimination to form the hydride complex and aldehyde is more favorable than the reverse, migratory insertion of C=O into the M–H bond. Since the unsaturated byproduct is thrown away, it's desirable to make it something small, cheap, and gaseous. Hydroxycarbonyl, formate, and tert-butyl ligands have been applied successfully with this goal in mind.

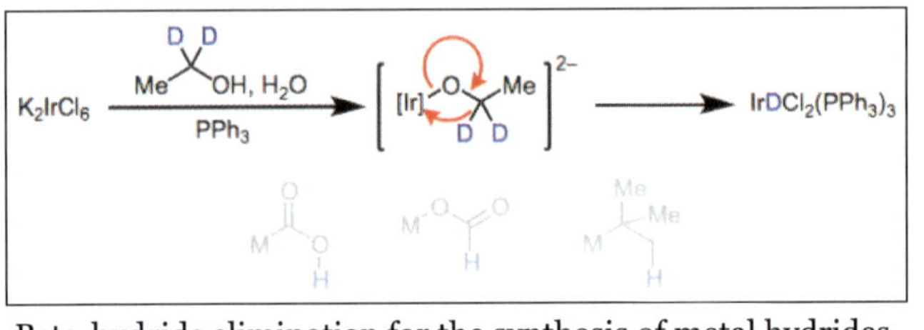

Beta-hydride elimination for the synthesis of metal hydrides.

Reactions

Metal hydrides are characterized by nucleophilic, electrophilic, and radical behavior. The exact behavior of a given metal hydride complex depends on its electronic properties, its M–H bond dissociation energy, and the nature of the reacting partner. Basic metal hydrides react with acids to free up a coordination site on the metal center (file this reaction alongside photolytic M–CO cleavage and M–CO cleavage with amine oxides).

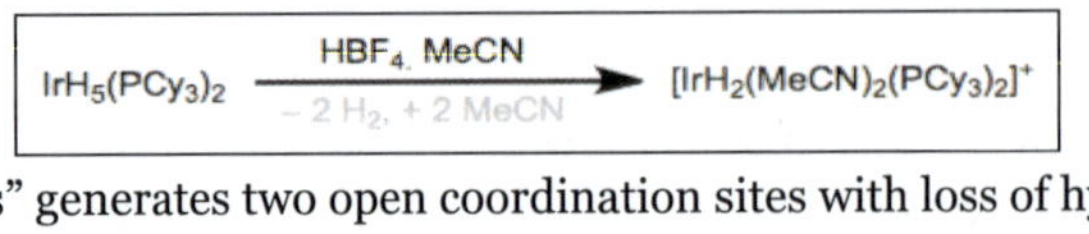

"Protonolysis" generates two open coordination sites with loss of hydrogen gas.

Migratory insertion reactions involving M–H bonds are extremely important in a practical sense and are conceptually related to nucleophilic hydride transfers from the metal center. We can think of insertions of π bonds into M–H as internal nucleophilic attack by the hydride ligand at one end of the π bond, with coordination of the metal center to the other. The transition state for migratory insertion of an olefin into M–H and an example reaction. Notice the coplanarity of M, H, and the C=C bond in the transition state for insertion, which determines the cis configuration of the product. The same is required for the microscopic reverse (β-hydride elimination) but this essential geometry is easily overlooked for β-hydride elimination.

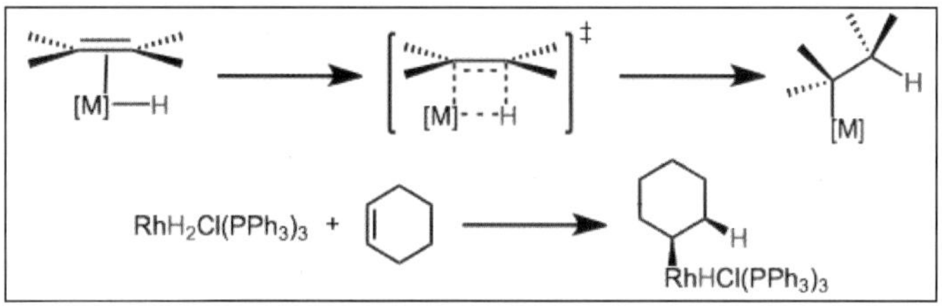

Migratory insertion of an olefin into M–H.
The relative configuration of M–C and C–H.

It is important to note that insertions of CO into M–H bonds are rare, because such insertions are usually unfavorable thermodynamically.

Finally, the radical behavior of certain metal hydrides in the presence of olefins deserves a nod. The behavior of the cobalt hydride complex in the reaction below is typical of these types of reactions.

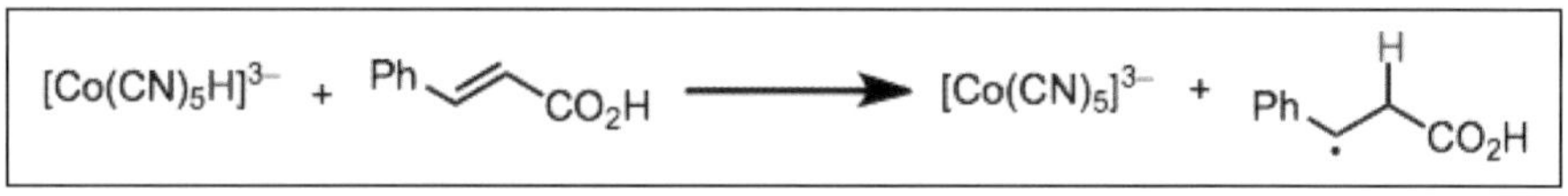

Hydrogen atom transfer to olefins. Radical reduction
of carbon tetrachloride is a related process.

And there you have it: the properties, synthesis, and reactivity of metal hydrides in a nutshell. Like other ligands we've seen so far, the behavior of hydrides is controlled by the nature of the metal center and its accompanying ligands. However, it's interesting to note that the observed behavior of hydrides often depends on the nature of the reaction conditions, as well. Some complexes display "schizophrenic" behavior, so to speak, putting on their nucleophile hat in the presence of an electrophile and their electrophile hat in the presence of a nucleophile.

References

- What-are-inorganic-compounds-definition-characteristics-examples: study.com, Retrieved 17 March, 2019
- Inorganic-compounds-essential-to-human-functioning: lumenlearning.com, Retrieved 14 June, 2019
- Introduction-to-acids, acids-bases-and-salts, chemistry: toppr.com, Retrieved 4 March, 2019
- Definition-of-acid-and-examples: thoughtco.com, Retrieved 2 July, 2019
- Introduction-to-bases, acids-bases-and-salts, chemistry: toppr.com, Retrieved 11 January, 2019
- More-about-salts, acids-bases-and-salts, chemistry: toppr.com, Retrieved 28 April, 2019
- Oxides, Compounds, Main-Group-Reactions, Descriptive-Chemistry, Inorganic-Chemistry: libretexts.org, Retrieved 25 May, 2019
- Metal-hydrides: thebalance.com, Retrieved 5 February, 2019
- Metal-Hydrides, Organometallic-Ligands, Inorganic-Chemistry: libretexts.org, Retrieved 20 May, 2019

4

Coordination Compounds

A compound which contains coordinate bonds is known as a coordination compound. Some of the different properties exhibited by coordination compounds are denticity, hapticity and chirality. This chapter has been carefully written to provide an easy understanding of these varied facets of coordination compounds.

Coordination chemistry is the study of compounds that have a central atom (often metallic) surrounded by molecules or anions, known as ligands. The ligands are attached to the central atom by dative bonds, also known as coordinate bonds, in which both electrons in the bond are supplied by the same atom on the ligand.

Coordination compound is any of a class of substances with chemical structures in which a central metal atom is surrounded by nonmetal atoms or groups of atoms, called ligands, joined to it by chemical bonds. Coordination compounds include such substances as vitamin B_{12}, hemoglobin, and chlorophyll, dyes and pigments, and catalysts used in preparing organic substances.

A major application of coordination compounds is their use as catalysts, which serve to alter the rate of chemical reactions. Certain complex metal catalysts, for example, play a key role in the production of polyethylene and polypropylene. In addition, a very stable class of organometallic coordination compounds has provided impetus to the development of organometallic chemistry. Organometallic coordination compounds are sometimes characterized by "sandwich" structures, in which two molecules of an unsaturated cyclic hydrocarbon, which lacks one or more hydrogen atoms, bond on either side of a metal atom. This results in a highly stable aromatic system.

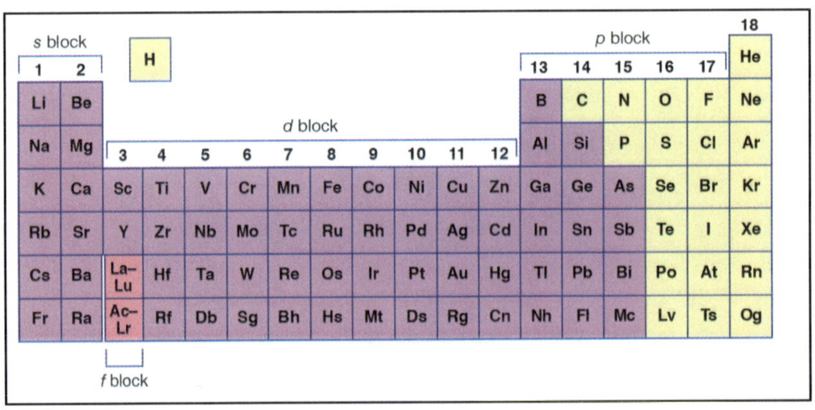

Organometallic coordination compounds, which include transition metal compounds, may be characterized by "sandwich" structures that contain two unsaturated cyclic hydrocarbons on either side of a metal atom. Organometallic compounds are found in the p-, d-, s-, and f- blocks of the periodic table (the purple-shaded blocks; the transition metals include those elements in the d- and f-blocks).

Naturally occurring coordination compounds are vital to living organisms. Metal complexes play a variety of important roles in biological systems. Many enzymes, the naturally occurring catalysts that regulate biological processes, are metal complexes (metalloenzymes); for example, carboxypeptidase, a hydrolytic enzyme important in digestion, contains a zinc ion coordinated to several amino acid residues of the protein. Another enzyme, catalase, which is an efficient catalyst for the decomposition of hydrogen peroxide, contains iron-porphyrin complexes. In both cases, the coordinated metal ions are probably the sites of catalytic activity. Hemoglobin also contains iron-porphyrin complexes, its role as an oxygen carrier being related to the ability of the iron atoms to coordinate oxygen molecules reversibly. Other biologically important coordination compounds include chlorophyll (a magnesium-porphyrin complex) and vitamin B_{12}, a complex of cobalt with a macrocyclic ligand known as Corrine.

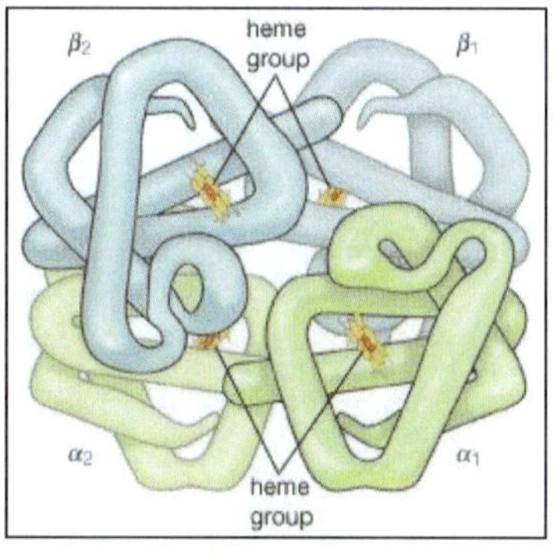

Hemoglobin is a protein made up of four polypeptide chains (α_1, α_2, β_1, and β_2). Each chain is attached to a heme group composed of porphyrin (an organic ringlike compound) attached to an iron atom. These iron-porphyrin complexes coordinate oxygen molecules reversibly, an ability directly related to the role of hemoglobin in oxygen transport in the blood.

The applications of coordination compounds in chemistry and technology are many and varied. The brilliant and intense colors of many coordination compounds, such as Prussian blue, render them of great value as dyes and pigments. Phthalocyanine complexes (e.g., copper phthalocyanine), containing large-ring ligands closely related to the porphyrins, constitute an important class of dyes for fabrics.

Several important hydrometallurgical processes utilize metal complexes. Nickel, cobalt, and copper can be extracted from their ores as ammine complexes using aqueous ammonia. Differences in the stabilities and solubility of the ammine complexes can be utilized in selective precipitation procedures that bring about separation of the metals. The purification of nickel can be effected by reaction with carbon monoxide to form the volatile tetracarbonylnickel complex, which can be distilled and thermally decomposed to deposit the pure metal. Aqueous cyanide solutions usually are employed to separate gold from its ores in the form of the extremely stable dicyanoaurate(−1) complex. Cyanide complexes also find application in electroplating.

There are a number of ways in which coordination compounds are used in the analysis of various substances. These include:

1. The selective precipitation of metal ions as complexes—for example, nickel(2+) ion as the dimethylglyoxime complex;

2. The formation of colored complexes, such as the tetrachlorocobaltate(2−) ion, which can be determined spectrophotometrically—that is, by means of their light absorption properties;

3. The preparation of complexes, such as metal acetylacetonates, which can be separated from aqueous solution by extraction with organic solvents.

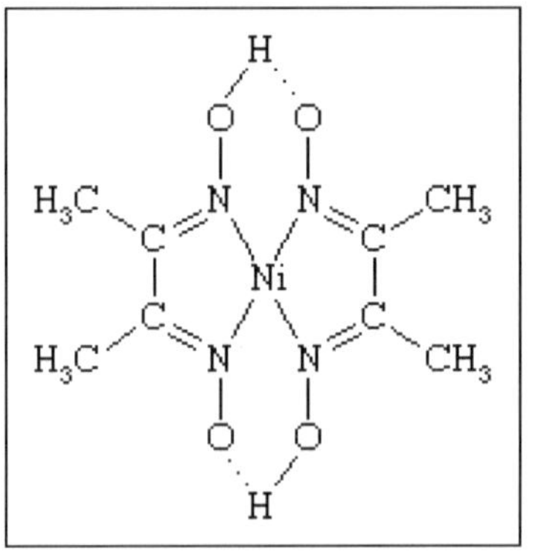

In certain circumstances, the presence of metal ions is undesirable, as, for example, in water, in which calcium (Ca^{2+}) and magnesium (Mg^{2+}) ions cause hardness. In such cases the undesirable effects of the metal ions frequently can be eliminated by "sequestering" the ions as harmless complexes through the addition of an appropriate complexing reagent. Ethylenediaminetetraacetic acid(EDTA) forms very stable complexes, and it is widely used for this purpose. Its applications include water softening (by tying up $Ca^{2+}$and Mg^{2+}) and the preservation of organic substances, such as vegetable oils and rubber, in which case it combines with traces of transition metal ions that would catalyze oxidation of the organic substances.

A technological and scientific development of major significance was the discovery in 1954 that certain complex metal catalysts—namely, a combination of titanium trichloride, or $TiCl_3$, and triethylaluminum, or $Al(C_2H_5)_3$—bring about the polymerizations of organic compounds with carbon-carbon double bonds under mild conditions to form polymers of high molecular weight and highly ordered (stereoregular) structures. Certain of these polymers are of great commercial importance because they are used to make many kinds of fibres, films, and plastics. Other technologically important processes based on metal complex catalysts include the catalysis by metal carbonyls, such as hydridotetracarbonylcobalt, of the so-called hydroformylation of olefins—i.e., of their reactions with hydrogen and carbon monoxide to form aldehydes—and the catalysis by tetrachloropalladate(2−) ions of the oxidation of ethylene in aqueous solution to acetaldehyde.

Characteristics of Coordination Compounds

Coordination compounds have been studied extensively because of what they reveal about molecular structure and chemical bonding, as well as because of the unusual chemical nature and useful properties of certain coordination compounds. The general class of coordination compounds—or complexes, as they are sometimes called—is extensive and diverse. The substances in the class may be composed of electrically neutral molecules or of positively or negatively charged species (ions).

Among the many coordination compounds having neutral molecules is uranium(+6) fluoride, or uranium hexafluoride (UF_6). The structural formula of the compoundrepresents the actual arrangement of atoms in the molecules:

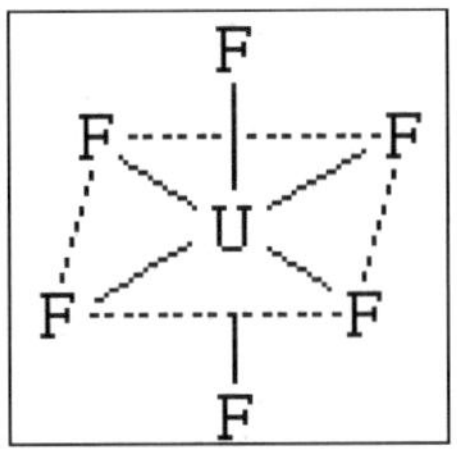

In this formula the solid lines, which represent bonds between atoms, show that four of the fluorine (F) atoms are bonded to the single atom of uranium (U) and lie in a plane with it, the plane being indicated by dotted lines (which do not represent bonds), whereas the remaining two fluorine atoms (also bonded to the uranium atom) lie above and below the plane, respectively.

An example of an ionic coordination complex is the hydrated ion of nickel, (Ni), hexaaquanickel(2+) ion, $[Ni(H_2O)_6]^{2+}$, the structure of which is. In this structure, the symbols and lines are used as above, and the brackets and the "two plus" (2+) sign show that the double positive charge is assigned to the unit as a whole.

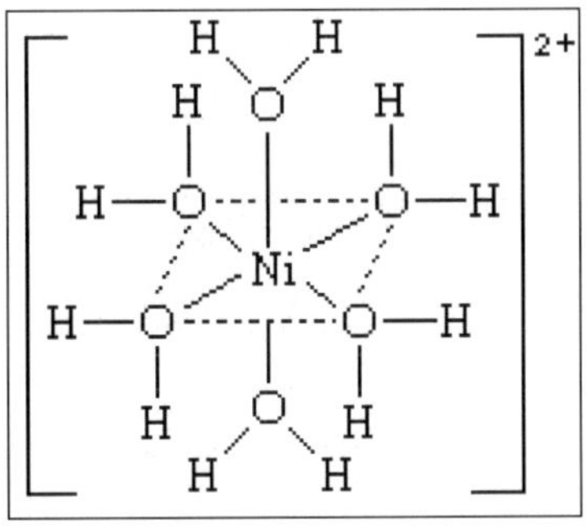

The central metal atom in a coordination compound itself may be neutral or charged (ionic). The coordinated groups—or ligands—may be neutral molecules such as water, ammonia (NH_3), or carbon monoxide (CO); negatively charged ions (anions) such as the fluoride (in the first example above) or cyanideion (CN^-); or, occasionally, positively charged ions (cations) such as the hydrazinium ($N_2H_5^+$) or nitrosonium (NO^+) ion.

Complex ions—that is, the ionic members of the family of coordination substances—may exist as free ions in solution, or they may be incorporated into crystalline materials (salts) with other ions of opposite charge. In such salts, the complex ion may be either the cationic (positively charged) or the anionic (negatively charged) component (or, on occasion, both). The hydrated nickel ion (above) is an example of a cationic complex. An anionic complex is the hexacyanide of the ferric iron (Fe^{+3}) ion, the hexacyanoferrate(3−) ion, $[Fe(CN)_6]^{3-}$, or Crystalline salts containing complex ions include potassium hexacyanoferrate(3−) (potassium ferricyanide), $K_3[Fe(CN)_6]$, and the hexahydrate of nickel chloride, hexaaquanickel(2+) chloride, $[Ni(H_2O)_6]Cl_2$. In each case the charge on the complex ion is balanced by ions of opposite charge. In the case of potassium ferricyanide, three positively charged potassium ions, K^+, balance the negative charge on the complex, and in the nickel complex the positive charges are balanced by two negative chloride ions, Cl^-. The oxidation state of the central metal is determined from the charges on the ligands and the overall charge on the complex. For example, in hexaaquanickel(2+), water is electrically neutral and the charge on the complex ion is +2; thus, the oxidation state of Ni is +2. In hexacyanoferrate(3−), all six cyano ligands have a charge of −1; thus, the overall charge of −3 dictates that the oxidation state of Fe is +3.

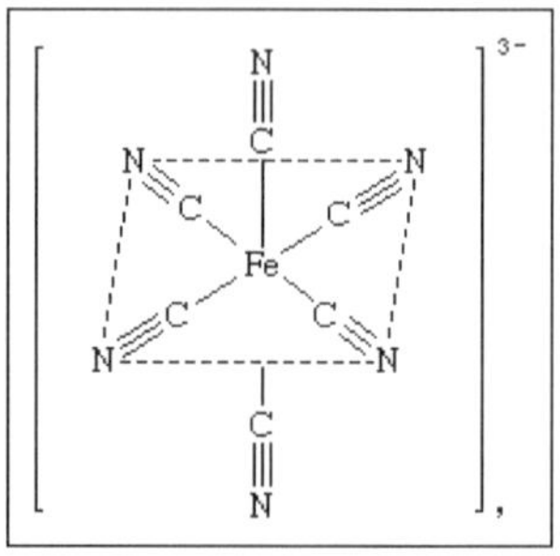

The distinction between coordination compounds and other substances is, in fact, somewhat arbitrary. The designationcoordination compound, however, is generally restricted to substances whose molecules or ions are discrete entities and in which the central atom is metal. Accordingly, molecules such as sulfur(+6) fluoride (sulfur hexafluoride; SF_6) and carbon(+4) fluoride (carbon tetrafluoride; CF_4) are not normally considered coordination compounds, because sulfur (S) and carbon (C) are nonmetallic elements. Yet there is no great difference between these compounds and, say, uranium hexafluoride. Furthermore, such simple ionic salts as sodium chloride (NaCl) or nickel(+2) fluoride (nickel difluoride; NiF_2) are not considered coordination compounds, because they consist of continuous ionic lattices rather than discrete molecules. Nevertheless, the arrangement (and bonding) of the anions surrounding the metal ions in these salts is similar to that in coordination compounds. Coordination compounds generally display a variety of distinctive physical and chemical properties, such as color, magnetic susceptibility, solubility and volatility, an ability to undergo oxidation-reduction reactions, and catalytic activity.

A coordination compound is characterized by the nature of the central metal atom or ion, the oxidation state of the latter (that is, the gain or loss of electrons in passing from the neutral atom to the charged ion, sometimes referred to as the oxidation number), and the number, kind, and arrangement of the ligands. Because virtually all metallic elements form coordination compounds—sometimes in several oxidation states and usually with many different kinds of ligands—a large number of coordination compounds are known.

Coordination Number

Coordination number is the term proposed by Werner to denote the total number of bonds from the ligands to the metal atom. Coordination numbers generally range between 2 and 12, with 4 (tetracoordinate) and 6 (hexacoordinate) being the most common. Werner referred to the central atom and the ligands surrounding it as the coordination sphere. Coordination number should be distinguished from oxidation number. The oxidation number, designated by an Arabic number with an appropriate sign (or, sometimes, by a Roman numeral in parentheses), is an index derived from a simple and formal set of rules and is not a direct indicator of electron distribution or of the charge on the central metal ion or compound as a whole. For the hexaamminecobalt (3+) ion, $[Co(NH_3)_6]^{3+}$, and the neutral moleculetriamminetrinitrocobalt(3+), $[Co(NO_2)_3(NH_3)_3]$, the coordination number of cobalt is 6 while its oxidation number is +3.

Ligands and Chelates

Each molecule or ion of a coordination compound includes a number of ligands, and, in any given substance, the ligands may be all alike, or they may be different. The term *ligand* was proposed by the German chemist Alfred Stock in 1916. Attachment of the ligands to the metal atom may be through only one atom, or it may be through several atoms. When only one atom is involved, the ligand is said to be monodentate; when two are involved, it is didentate, and so on. In general, ligands utilizing more than one bond are said to be polydentate. Because a polydentate ligand is joined to the metal atom in more than one place, the resulting complex is said to be cyclic—i.e., to contain a ring of atoms. Coordination compoundscontaining polydentate ligands are called chelates (from Greek *chele*, "claw"), and their formation is termed *chelation*. Chelates are particularly stable and useful. An example of a typical chelate is bis(1,2-ethanediamine)copper(2+), the complex formed between the cupric ion (Cu^{2+}) and the organic compound ethylenediamine($NH_2CH_2CH_2NH_2$, often abbreviated as en in formulas). The formula of the complex is $[Cu(NH_2CH_2CH_2NH_2)_2]^{2+}$ and the structural formula is,

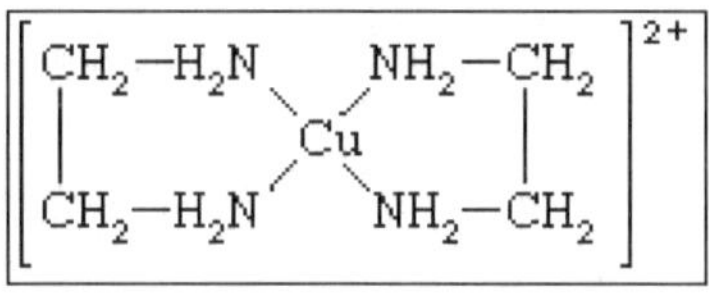

Mononuclear and Monodentate

The simplest types of coordination compounds are those containing a single metal atom or ion (mononuclear compounds) surrounded by monodentate ligands. Most of the coordination compounds already cited belong to this class. Among the ligands forming such complexes are a wide variety of neutral molecules (such as ammonia, water, carbon monoxide, and nitrogen), as well as monoatomic and polyatomic anions (such as the hydride, fluoride, chloride, oxide, hydroxide, nitrite, thiocyanate, carbonate, sulfate, and phosphate ions). Coordination of such ligands to the metal virtually always occurs through an atom possessing an unshared pair of electrons, which it donates to the metal to form a coordinate bond with the latter. Among the atoms that are known to coordinate to metals are those of virtually all the nonmetallic elements (such as hydrogen, carbon, oxygen, nitrogen, and sulfur), with the exception of the noble gases (helium [He], neon [Ne], argon [Ar], krypton [Kr], and xenon [Xe]).

Polydentate

The chelate complex of a copper ion and ethylenediamine is an example of a compound formed between a metal ion and a didentate ligand.

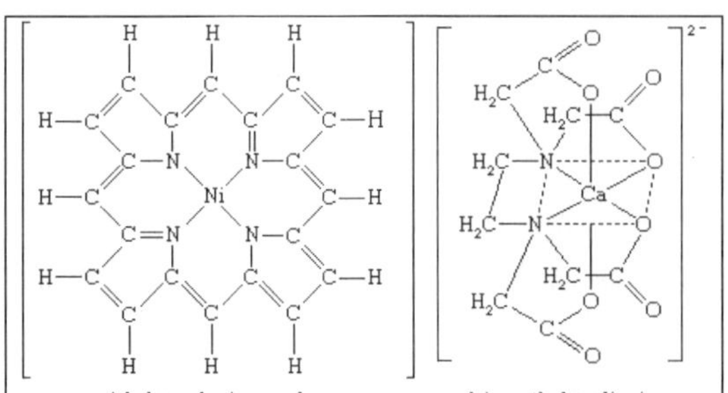

nickel-porphyrin complex

calcium-ethylenediamine-tetraacetate complex

These are a nickel complex with a tetradentate large-ring ligand, known as a porphyrin, and a calcium complex with a hexadentate ligand, ethylenediaminetetraacetate (EDTA). Because metal-ligand attachment in such chelate complexes is through several bonds, such complexes tend to be very stable.

The commonest and most stable complexes of the lanthanoid metals (the series of 14 *f*-block elements following lanthanum [atomic number 57]) are those with chelating oxygen ligands, such as EDTA-type anions or hydroxo acids (e.g., tartaric or citric acids). The formation of such water-soluble complexes is employed in the separation of lanthanoids by ion-exchange chromatography. Lanthanoid β-diketonates are well known because some fluorinated β-diketonates yield volatile complexes amenable to gas-chromatographic separations. Neutral complexes can complex further to yield anionic species such as octacoordinated tetrakis (thenoyltrifluoroacetato) neodymate(1–), $[Nd(CF_3COCHCOCF_3)_4]^-$.

Certain ligands may be either monodentate or polydentate, depending on the particular compound in which they occur. The carbonate ion, $(CO_3)^{2-}$, for example, is coordinated to the cobalt (Co^{3+}) ions in two cobalt complexes, pentaamminecarbonatocobalt(+), $[Co(CO_3)(NH_3)_5]^+$, and tetraamminecarbonatocobalt(+), $[Co(CO_3)(NH_3)_4]^+$, through one and two oxygen atoms, respectively.

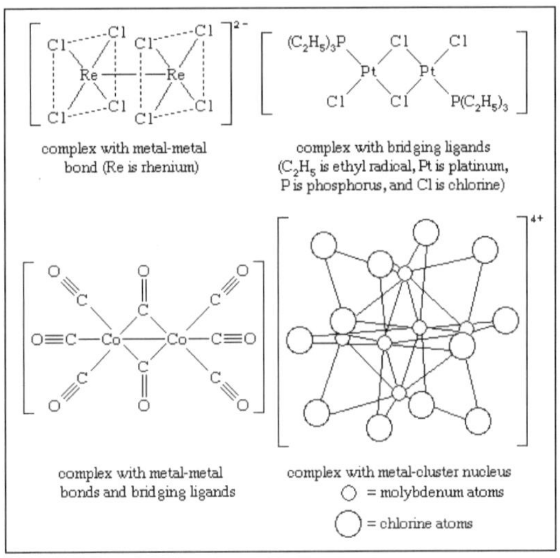

complex with metal-metal bond (Re is rhenium)

complex with bridging ligands (C_2H_5 is ethyl radical, Pt is platinum, P is phosphorus, and Cl is chlorine)

complex with metal-metal bonds and bridging ligands

complex with metal-cluster nucleus
○ = molybdenum atoms
○ = chlorine atoms

Polynuclear

Polynuclear complexes are coordination compounds containing two or more metal atoms, or ions, in a single coordination sphere. The two atoms may be held together through direct metal-metal bonds, through bridging ligands, or both. Examples of each are shown above, along with a unique metal-cluster complex having six metal atoms in its nucleus.

Nomenclature

Generally, the systematic naming of coordination compounds is carried out by rules recommended by the International Union of Pure and Applied Chemistry (IUPAC). Among the more important of these are the following:

1. Neutral and cationic complexes are named by first identifying the ligands, followed by the metal; its oxidation number may be given in Roman numerals enclosed within parentheses. Alternatively,

the overall charge on the complex may be given in Arabic numbers in parentheses. This convention is generally followed here. In formulas, anionic ligands (ending in *-o*; in general, if the anion name ends in *-ide*, *-ite*, or *-ate*, the final *e* is replaced by *-o*, giving *-ido*, *-ito*, and *-ato*) are cited in alphabetical order ahead of neutral ones also in alphabetical order (multiplicative prefixes are ignored). When the complex contains more than one ligand of a given kind, the number of such ligands is designated by one of the prefixes *di-*, *tri-*, *tetra-*, *penta-*, and so on or, in the case of complex ligands, by *bis-*, *tris-*, *tetrakis*, *pentakis-*, and so on. In names (as opposed to formulas) the ligands are given in alphabetical order without regard to charge. The oxidation number of the metal is defined in the customary way as the residual charge on the metal if all the ligands were removed together with the electron pairs involved in coordination to the metal. The following examples are illustrative (aqua is the name of the water ligand):

$[Ce(CH_3CCHO_2CCH_3)_4]$	tetrakis (2,4-pentanedionato cerium(IV)
$[Nd(H_2O)_9]^{3+}$	nonaaquanedymium(III)

2. Anionic complexes are similarly named, except that the name is terminated by the suffix *-ate*; for example:

$[Er(NCS)_6]^{3-}$	hexakis(thiocyanato)erbate(III)

3. In the case of salts, the cation is named first and then the anion; for example:

$(NH_4)_2[Ce(NO_3)_6]$	ammonium hexanitatoeerate(IV)

4. Polynuclear complexes are named as follows, bridging ligands being identified by a prefix consisting of the Greek letter *mu* (μ-):

$[(CO)_5Mn - Mn(CO)_5]$	decacarbonyldi-manganese(0)or bis(pentacarbonyl-manganese)
$[(NH_3)_5Cr - OH - Cr(NH_3)_5]Cl_5$	μ – hudroxo-bis[penta-ammine-chormium(III)]chloride

In addition to their systematic designations, many coordination compounds are also known by names reflecting their discoverers or colors. Examples are:

$K[PtCl_3(C_2H_4)] \cdot H_2O$	Zeise's salt
$NH_4[Cr(SCN)_4(NH_3)_2] \cdot H_2O$	Reinecke's salt
$[Co(NH_3)_5(H_2O)]Cl_3$	Roseocobaltic chloride(red)
$KFe[Fe(CN)_6]$	Prussian blue

Structure and Bonding of Coordination Compounds

Werner originally postulated that coordination compounds can be formed because the central atoms carry the capacity to form secondary, or coordinate, bonds, in addition to the normal, or valence, bonds. A more complete description of coordinate bonding, in terms of electron pairs, became possible in the 1920s, following the introduction of the concept that all covalent bonds consist of electron pairs shared between atoms, an idea advanced chiefly by the American physical chemist Gilbert N. Lewis. In Lewis's formulation, when both electrons are contributed by one of the atoms, as in the boron-nitrogen bond formed when the substance boron trifluoride (BF_3) combines with ammonia, the bond is called a coordinate bond.

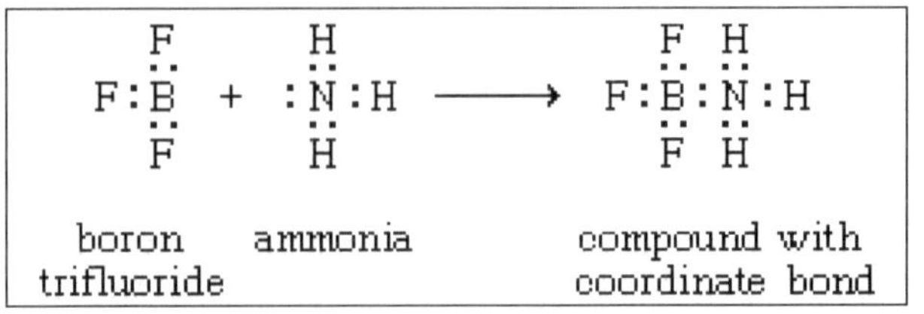

In Lewis's formulas, the valence (or bonding) electrons are indicated by dots, with each pair of dots between two atomic symbols representing a bond between the corresponding atoms.

Following Lewis's ideas, the suggestion was made that the bonds between metals and ligands were of this same type, with the ligands acting as electron donors and the metal ions as electron acceptors. This suggestion provided the first electronic interpretation of bonding in coordination compounds. The coordination reaction between silver ions and ammonia illustrates the resemblance of coordination compounds to the situation in the boron-nitrogen compound. According to this view, the metal ion can be regarded as a so-called Lewis acid and the ligands as Lewis bases.

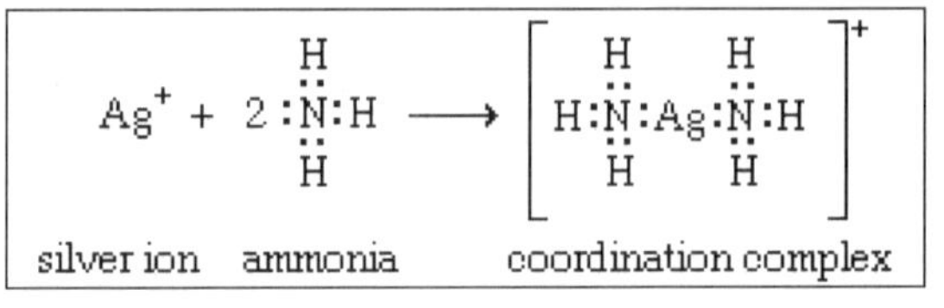

A coordinate bond may also be denoted by an arrow pointing from the donor to the acceptor.

Geometry

Many coordination compounds have distinct geometric structures. Two common forms are the square planar, in which four ligands are arranged at the corners of a hypothetical square around the central metal atom, and the octahedral, in which six ligands are arranged, four in a plane and one each above and below the plane. Altering the position of the ligands relative to one another can produce different compounds with the same chemical formula. Thus, a cobaltion linked to two chloride ions and four molecules of ammonia can occur in both green and violet forms according to how the six ligands are placed. Replacing a ligand also can affect the color. A cobalt ion linked to six ammonia molecules is yellow. Replacing one of the ammonia molecules with a water molecule turns it rose red. Replacing all six ammonia molecules with water molecules turns it purple.

Among the essential properties of coordination compounds are the number and arrangement of the ligands attached to the central metal atom or ion—that is, the coordination number and the coordination geometry, respectively. The coordination number of a particular complex is determined by the relative sizes of the metal atom and the ligands, by spatial (steric) constraints governing the shapes (conformations) of polydentate ligands, and by electronic factors, most notably the electronic configuration of the metal ion. Although coordination numbers from 1 to 16 are known, those below 3 and above 8 are rare. Possible structures and examples of species for the various coordination numbers are as follows: three, trigonal planar ($[Au\{P(C_6H_5)_3\}_3]^+$;

four, tetrahedral ($[CoCl_4]^{2-}$) or square planar ($[PtCl_4]^{2-}$); five, trigonal bipyramid ([$CuCl_5$]>}]$^{3-}$) or square pyramid ($VO(acetylacetonate)_2$); six, octahedral ($[Co(NO_2)_6]^{3-}$) or trigonal prismatic ([Re {$S_2C_2(C_6H_5)_2$}$_3$]); seven, pentagonal bipyramid ($Na_5[Mo(CN)_7]\cdot 10H_2O$), capped trigonal prism (cation in $[Ca(H_2O)_7]_2[Cd_6Cl_{16}(H_2O)_2]\cdot H_2O$), or capped octahedron (cation in $[Mo(CNC_6H_5)_7][PF_6]_2$); eight, square antiprism or dodecahedron ([$Zr(acetylacetonate)_4$]; and nine, capped square antiprism ($La(NH_3)_9]^{3+}$) or tricapped trigonal prism ($[ReH_9]^{2-}$).

Coordination numbers are also affected by the 18-electron rule (sometimes called the noble gas rule), which states that coordination compounds in which the total number of valence electrons approaches but does not exceed 18 (the number of electrons in the valence shells of the noble gases) are most stable. The stabilities of 18-electron valence shells are also reflected in the coordination numbers of the stable mononuclear carbonyls of different metals that have oxidation number 0—e.g., tetracarbonylnickel, pentacarbonyliron, and hexacarbonylchromium (each of which has a valence shell of 18).

The 18-electron rule applies particularly to covalent complexes, such as the cyanides, carbonyls, and phosphines. For more ionic (also called outer-orbital) complexes, such as fluoro or aqua complexes, electronic factors are less important in determining coordination numbers, and configurations corresponding to more than 18 valence electrons are not uncommon. Several nickel(+2) complexes, for example—including the hexafluoro, hexaaqua, and hexaammine complexes—each have 20 valence electrons.

Any one metal ion tends to have the same coordination number in different complexes—e.g., generally six for chromium(+3)—but this is not invariably so. Differences in coordination number may result from differences in the sizes of the ligands; for example, the iron(+3) ion is able to accommodate six fluoride ions in the hexafluoro complex [FeF_6]>]$^{3-}$ but only four of the larger chloride ions in the tetrachloro complex $[FeCl_4]^-$. In some cases, a metal ion and a ligand form two or more complexes with different coordination numbers—e.g., tetracyanonickelate [$Ni(CN)_4$]>]$^{2-}$ and pentacyanonickelate [$Ni(CN)_5$]>]$^{3-}$, both of which contain Ni in the +2 oxidation state.

Isomerism

Coordination compounds often exist as isomers—i.e., as compounds with the same chemical composition but different structural formulas. Many different kinds of isomerism occur among coordination compounds. The following are some of the more common types.

Cis-trans Isomerism

Cis-trans (geometric) isomers of coordination compounds differ from one another only in the manner in which the ligands are distributed spatially; for example, in the isomeric pair of diamminedichloroplatinum compounds the two ammonia molecules and the two chlorine atoms are situated next to one another in one isomer, called the cis (Latin for "on this side") isomer, and across from one another in the other, the trans (Latin for "on the other side") isomer.

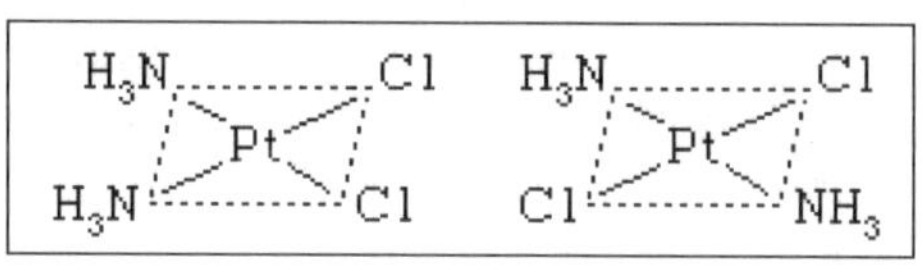

A similar relationship exists between the *cis* and *trans* forms of the tetraamminedichlorocobalt(1+) ion:

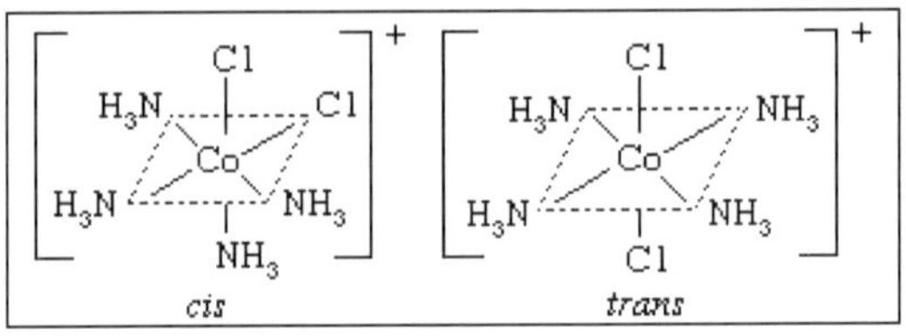

cis *trans*

Enantiomers and Diastereomers

So-called optical isomers (or enantiomers) have the ability to rotate plane-polarized light in opposite directions. Enantiomers exist when the molecules of the substances are mirror images but are not superimposable upon one another. In coordination compounds, enantiomers can arise either from the presence of an asymmetric ligand, such as one isomer of the amino acid, alanine (aminopropionic acid), (structure below) or from an asymmetric arrangement of the ligands.

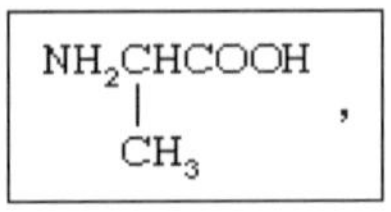

Familiar examples of the latter variety are octahedral complexes carrying three didentate ligands, such as ethylenediamine, $NH_2CH_2CH_2NH_2$. The two enantiomers corresponding to such a complex are depicted by the structures below.

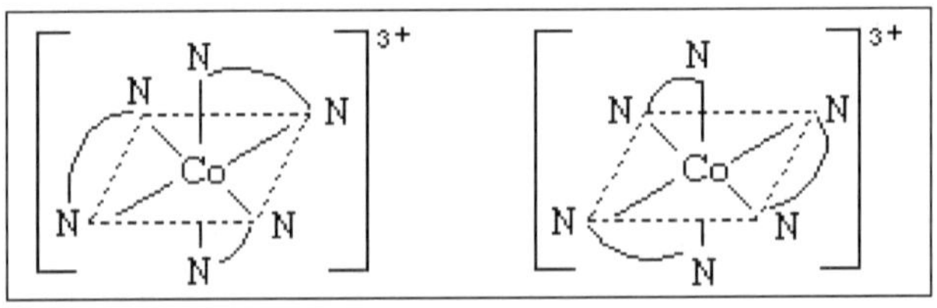

The ethylenediamine ligands above are indicated by a curved line between the symbols for the nitrogen atoms.

Diastereomers, on the other hand, are not superimposable and also are not mirror images. Using AB as an example of a chelating ligand, in which the symbol AB implies that the two ends of the chelate are different, there are six possible isomers of a complex *cis*-$[M(AB)_2X_2]$. For example, AB might correspond to alanine $[CH_3CH(NH_2)C(O)O]^-$, where both N and O are attached to the metal. Alternatively, AB could represent a ligand such as propylenenediamine, $[NH_2CH_2C(CH_3)HNH_2]$, where the two ends of the molecule are distinguished by the fact that one of the Hs on a C is substituted with a methyl (CH_3) group.

Ionization Isomerism

Certain isomeric pairs occur that differ only in that two ionic groups exchange positions within (and without) the primary coordination sphere. These are called ionization isomers and are exemplified by the two compounds, pentaamminebromocobalt sulfate, $[CoBr(NH_3)_5]SO_4$, and pentaamminesulfatocobalt bromide, $[Co(SO_4)(NH_3)_5]Br$. In the former the bromide ion is coordinated to the cobalt(3+) ion, and the sulfate ion is outside the coordination sphere; in the latter the sulfate ion occurs within the coordination sphere, and the bromide ion is outside it.

Linkage Isomerism

Isomerism also results when a given ligand is joined to the central atom through different atoms of the ligand. Such isomerism is called linkage isomerism. A pair of linkage isomers are the ions $[Co(NO_2)(NH_3)_5]^{2+}$and $[Co(ONO)(NH_3)_5]^{2+}$, in which the anionic ligand is joined to the cobalt atom through nitrogen or oxygen, as shown by designating it with the formulas NO_2^-(nitro) and ONO^- (nitrito), respectively. Another example of this variety of isomerism is given by the pair of ions $[Co(CN)_5(NCS)]^{3-}$ and $[Co(CN)_5(SCN)]^{3-}$, in which an isothiocyanate $(NCS)^-$ and a thiocyanate group $(SCN)^-$ are bonded to the cobalt(3+) ion through a nitrogen or sulfur atom, respectively.

Coordination Isomerism

Ionic coordination compounds that contain complex cations and anions can exist as isomers if the ligands associated with the two metal atoms are exchanged, as in the pair of compounds, hexaamminecobalt(3+) hexacyanochromate(3−), $[Co(NH_3)_6][Cr(CN)_6]$, and hexaamminechromium(3+) hexacyanocobaltate(3−), $[Cr(NH_3)_6][Co(CN)_6]$. Such compounds are called coordination isomers, as are the isomeric pairs obtained by redistributing the ligands between the two metal atoms, as in the doubly coordinated pair, tetraammineplatinum(2+) hexachloroplatinate(2−), $[Pt(NH_3)_4][PtCl_6]$, and tetraamminedichloroplatinum(2+) tetrachloroplatinate(2−), $[PtCl_2(NH_3)_4][PtCl_4]$.

Ligand Isomerism

Isomeric coordination compounds are known in which the overall isomerism results from isomerism solely within the ligand groups. An example of such isomerism is shown by the ions, bis(1,3-diaminopropane)platinum(2+) and bis(1,2-diaminopropane)platinum(2+).

$[Pt(NH_2CH_2CH_2CH_2NH_2)_2]^{2+}$ and $[Pt(NH_2CH(CH_3)CH_2NH_2)_2]^{2+}$.

Bonding Theories

Valence Bond Theory

Several theories currently are used to interpret bonding in coordination compounds. In the valence bond (VB) theory, proposed in large part by the American scientists Linus Pauling and John C. Slater, bonding is accounted for in terms of hybridized orbitals of the metal ion, which is assumed to possess a particular number of vacant orbitals available for coordinate bonding that equals its coordination number. Each ligand donates an electron pair to form a coordinate-covalent bond, which is formed by the overlap of an unoccupied orbital of the metal ion and a filled orbital of a ligand. The configuration of the complex depends on the type and number of orbitals involved in the hybridization—e.g. sp (linear), sp^3 (tetrahedral), dsp^2 (square planar), and d^2sp^3 (octahedral), in which the superscripts denote the number of orbitals of a particular type. In many cases, the number of unpaired electrons, as determined by magnetic susceptibility measurements, agrees with the theoretical prediction. The theory was modified in 1952 by the Canadian-born American Nobel chemistry laureate Henry Taube, who distinguished between inner orbital complexes (d^2sp^3) and outer orbital complexes (sp^3d^2) to account for discrepancies between octahedral complexes. The main defect of the simple VB theory lies in its failure to include the antibonding molecular orbitals produced during complex formation. Thus, it fails to offer an explanation for the striking colors of many complexes, which arise from their selective absorption of light of only certain wavelengths. From the early 1930s through the early 1950s, VB theory was used to interpret almost all coordination phenomena, for it gave simple answers to the questions of geometry and magnetic susceptibility with which chemists of that time were concerned.

Crystal Field Theory

Considerable success in understanding certain coordination compounds also has been achieved by treating them as examples of simple ionic or electrostatic bonding. The German theoretical physicist Walther Kossel's ionic model of 1916 was revitalized and developed by the American physicists Hans Bethe and John H. Van Vleck into the crystal field theory (CFT) of coordination, used by physicists as early as the 1930s but not generally accepted by chemists until the 1950s. This view attributes the bonding in coordination compounds to electrostatic forces between the positively charged metal ions and negatively charged ligands—or, in the case of neutral ligands (e.g., water and ammonia), to charge separations (dipoles) that appear within the molecules. Although this approach meets with considerable success for complexes of metal ions with small electronegative ligands, such as fluoride or chloride ions or water molecules, it breaks down for ligands of low polarity (charge separation), such as carbon monoxide. It also requires modification to explain why the spectral (light-absorption) and magnetic properties of coordinated metal ions generally differ from those of the free ions and why, for a given metal ion, these properties depend on the nature of the ligands.

Ligand Field and Molecular Orbital Theories

Since 1950 it has been apparent that a more complete theory, which incorporates contributions from both ionic and covalent bonding, is necessary to give an adequate account of the properties of coordination compounds. Such a theory is the so-called ligand field theory (LFT), which has its origin in the more general, but more complicated, theory of chemical bonding called the molecular orbital (MO) theory. (Molecular orbitals describe the spatial distributions of electrons in molecules, just as atomic orbitals describe the distributions in atoms.) This theory accounts with remarkable success for most properties of coordination compounds.

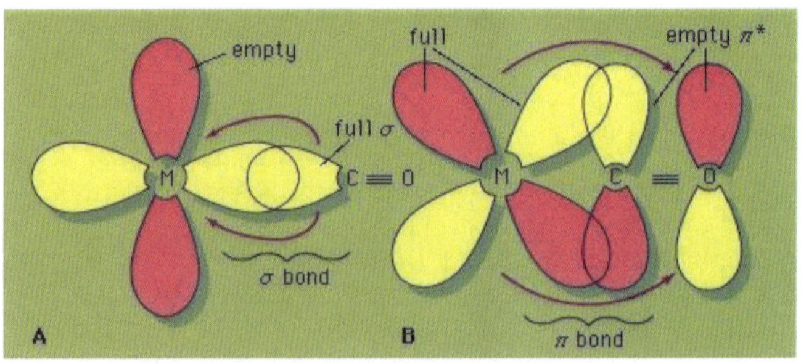

Carbon monoxide is a neutral ligand, meaning it does not carry an ionic charge.

The empty π orbitals in carbon monoxide molecules accept d orbital electrons from metal atoms, thereby stabilizing the oxidation state of metal atoms.

The magnetic properties of a coordination compound can provide indirect evidence of the orbital energy levels used in bonding. Hund rules, which describe the order in which electrons fill atomic shells, require that the maximum number of unpaired electrons in energy levels have equal or almost equal energies. Compounds that contain no unpaired electrons are slightly repelled by a magnetic field and are said to be diamagnetic. Because unpaired electrons behave like tiny magnets, compounds that contain unpaired electrons are attracted by a magnetic field and are said to be paramagnetic. The measure of a compound's magnetism is called its magnetic moment. The complex ion hexafluoroferrate(3–) (FeF_6^{3-}) has a magnetic moment to be expected from a substance with five unpaired electrons, as does the free iron(3+) ion (Fe^{3+}), whereas the magnetic moment of the closely related hexacyanoferrate(3–) ($[Fe(CN)_6]^{3-}$), which also contains Fe^{3+}, corresponds to only one unpaired electron.

LFT is able to account for this difference in magnetic properties. For octahedral complexes the electrons of the ligands fill all six bonding molecular orbitals, whereas any electrons from the metal cation occupy the nonbonding (*t2g*) and antibonding (*eg*) orbitals. The orbital splitting between the two sets of orbitals (*t2g* and *eg*) is designated as the orbital ligand field parameter, δ*o* (where *o* stands for octahedral). Ligands whose orbitals interact strongly with the metal cation's orbitals are called strong-field ligands. For such ligands the orbital splitting is between the *t2g* and *eg* orbitals, and consequently the δ*o* value is large. Ligands whose orbitals interact only weakly with the metal cation's orbitals are called weak-field ligands. For such ligands the orbital splitting is between the *t2g* and *eg* orbitals, and consequently the δ*o*value is small. For transition metal ions with electron configurations d^0 through d^3 and d^8 through d^{10}, only one configuration is possible, so the net spin of the electrons in the complex is the same for both strong-field and weak-field ligands. In contrast, for transition metal ions with electron configurations d^4 through d^7 (Fe^{3+} is d^5), both high-spin and low-spin states are possible depending on the ligand involved. Strong-field ligands, such as the cyanide ion, result in low-spin complexes, whereas weak-field ligands, such as the fluoride ion, result in high-spin complexes. Therefore, in the $[Fe(CN)_6]^{3-}$ ion, all five electrons occupy the *t2g* orbitals, resulting in a magnetic moment indicating one unpaired electron; in the $[FeF_6]^{3-}$ ion, three electrons occupy the *t2g* orbitals and two electrons occupy the *eg* orbitals, resulting in a magnetic moment indicating five unpaired electrons.

An important conclusion from LFT is that two types of bonds, called sigma (σ) bonds and pi (π) bonds, occur in coordination compounds just as they do in ordinary covalent (organic) compounds. The more usual of the two are σ bonds, which are symmetrical about the axis of the bond; π bonds, which are less common, are unsymmetrical with regard to the bond axis. In coordination compounds, π bonding may result from donation of electrons from ligands, such as fluorine or oxygen atoms, to empty *d* orbitals of the metal atoms. An example of this type of bonding occurs in the chromate ion, $(CrO_4)^{2-}$, in which the oxygen atoms donate electrons to the central chromium ion (Cr^{6+}). Alternatively, electrons from *d* orbitals of the metal atom may be donated to empty orbitals of the ligand. This is the case in the compound tetracarbonylnickel, $Ni(CO)_4$, in which empty π orbitals in the carbon monoxide molecules accept *d*-orbital electrons from the nickel atom.

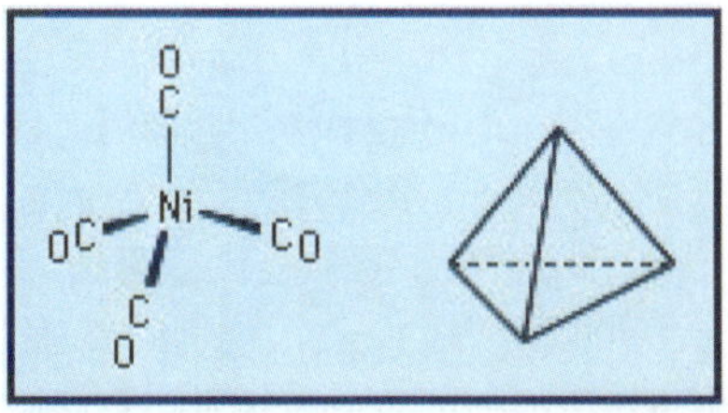

Empty π orbitals in carbon monoxide molecules accept d orbital electrons from nickel to form the compound tetracarbonylnickel, $Ni(CO)_4$.

Ligands may be classified according to their donor and acceptor abilities. Some ligands that possess no orbitals with symmetry appropriate for π bonding, such as ammonia, are σ donors only. On the other hand, ligands with occupied *p* orbitals are potential π donors and may donate these electrons along with the σ-bonding electrons. For ligands with vacant π^* or *d*orbitals, there is a possibility of π back bonding, and the ligands may be π acceptors. Ligands can be arranged in a so-called spectrochemical series in order from strong π acceptors (correlated with low spin, strong field, and large δ values) to strong π donors (correlated with high spin, weak field, and small δ values) as follows: CO, CN^- > 1, 10-phenanthroline > NO^{2-} > en > NH_3 > NCS^- > H_2O > F^- > $RCOO^-$ (where R is an alkyl group) > OH^- > Cl^- > Br^- > I^-. Additional ligands could be added here, but such an expanded list would not be very useful, because the order of the ligands is affected by the nature and charge on the metal ion, the presence of other ligands, and other factors.

The energy of the light absorbed as electrons are raised to higher levels is the difference in energy between the *d* orbital levels of transitional metal complexes. As a result, electronic spectra can provide direct evidence of orbital energy levels and information about bonding and electronic configurations in complexes. In some cases, these spectra can also provide information about the magnitude of the effect of ligands on the *d* orbitals of the metal (δ_0). The energy levels of *d*-electron configurations, as opposed to the energies of individual electrons, are complicated, since electrons in atomic orbitals can interact with each other. Tetrahedral complexes give more intense absorption spectra than do octahedral complexes. For *f*-orbital systems (lanthanoids, 4*f*n, and actinoids, 5*f*n) the LFT treatment is similar to that for *d*-orbital systems. However, the number of parameters is greater, and, even in complexes with cubic symmetry, two parameters are needed to describe the splittings of the *f* orbitals. Furthermore, *f*-orbital wave functions are not well known, and interpretation of the properties of *f*-electron systems is much more difficult than it is for *d* systems. In an effort to overcome such difficulties with *f*-orbital systems, an approach called the angular overlap model (AOM) was developed, but it proved of relatively little value for these systems.

Principal Types of Complexes

The tendency for complexes to form between a metal ion and a particular combination of ligands and the properties of the resulting complexes depend on a variety of properties of both the metal ion and the ligands. Among the pertinent properties of the metal ion are its size, charge, and electron configuration. Relevant properties of the ligand include its size and charge, the number and kinds of atoms available for coordination, the sizes of the resulting chelate rings formed (if any), and a variety of other geometric (steric) and electronic factors.

Many elements, notably certain metals, exhibit a range of oxidation states—that is, they are able to gain or lose varying numbers of electrons. The relative stabilities of these oxidation states are markedly affected by coordination of different ligands. The highest oxidation states correspond to empty or nearly empty *d* subshells (as the patterns of *d* orbitals are called). These states are generally stabilized most effectively by small negative ligands, such as fluorine and oxygen atoms, which possess unshared electron pairs. Such stabilization reflects, in part, the contribution of π bonding caused by electron donation from the ligands to empty *d*orbitals of the metal ions in the complexes. Conversely, neutral ligands, such as carbon monoxide and unsaturated hydrocarbons, which are relatively poor electron donors but which can accept π electrons from filled *d* orbitals of the metal, tend to stabilize the lowest oxidation states of metals. Intermediate oxidation states are most effectively stabilized by ligands such as water, ammonia, and cyanide ion, which are moderately good σ–electron donors but relatively poor π–electron donors or acceptors.

Chromium complexes of various oxidation states:

oxidation state	electron configuration*	coordination complex
+6	d^0	$[CrF_6]$, $[CrO_4]^{2-}$
+5	d^1	$[CrO_4]^{3-}$
+4	d^2	$[CrO_4]^{4-}$, $[Cr(OR)_4]$**
+3	d^3	$[Cr(H_2O)_6]^{3+}$, $[Cr(NH_3)_6]^{3+}$
+2	d^4	$[Cr(H_2O)_6]^{2+}$
0	d^6	$[Cr(CO)_6]$, $[Cr(C_6H_6)_2]$

*Number of d electrons indicated by superscript.
**R symbolizes an organic alkyl radical.

Aqua Complexes

Few ligands equal water with respect to the number and variety of metal ions with which they form complexes. Nearly all metallic elements form aqua complexes, frequently in more than one oxidation state. Such aqua complexes include hydrated ions in aqueous solution as well as hydrated salts such as hexaaquachromium(3+) chloride, $[Cr(H_2O)_6]Cl_3$. For metal ions with partially filled *d* subshells (i.e., transition metals), the coordination numbers and geometries of the hydrated ions in solution can be inferred from their light-absorption spectra, which are generally consistent with octahedral coordination by six water molecules. Higher coordination numbers probably occur for the hydrated rare-earth ions such as lanthanum(3+).

When other ligands are added to an aqueous solution of a metal ion, replacement of water molecules in the coordination sphere may occur, with the resultant formation of other complexes. Such replacement is generally a stepwise process, as illustrated by the following series of reactions that results from the progressive addition of ammonia to an aqueous solution of a nickel(2+) salt:

$$[Ni(H_2O)_6]^{2+} + NH_3 \rightleftharpoons [Ni(NH_3)(H_2O)_5]^{2+} + H_2O$$

With increasing additions of ammonia, the equilibria are shifted toward the higher ammine complexes (those with more ammonia and less water) until ultimately the hexaamminenickel(2+) ion predominates:

$$[Ni(NH_3)_5(H_2O)]^{2+} + NH_3 \rightleftharpoons [Ni(NH_3)_6]^{2+} + H_2O$$

The tendency of metal ions in aqueous solution to form complexes with ammonia as well as with organic amines (derivatives of ammonia, with chains of carbon atoms attached to the nitrogen atom) is widespread. The stabilities of such complexes exhibit a considerable range of dependence on the nature of the metal ion as well as on that of the amine. The marked enhancement of stability that results from chelation is reflected in the equilibrium constants of the reactions—values that indicate the relative proportions of the starting materials and the products at equilibrium.

Complexes of hexaaquanickel(2+) ions can be formed with a series of polyamines—i.e., in which L is the ligand and *n* the number of water molecules displaced from the complex.

$$[Ni(H_2O)_6]^{2+} + nL \rightleftharpoons [NiL_n(H_2O)_{6-n}]^{2+} + nH_2O$$

In this series the equilibrium constants, K_L, increase dramatically as the possibilities for chelation increase (that is, as the number of nitrogen atoms available for bonding to the metal atomincreases).

It should be noted that, in the particular examples cited above, the coordination number of the metal ion is invariant throughout the substitution process, but this is not always the case. Thus, the ultimate products of the addition of the cyanide ion to an aqueous solution of hexaaquanickel(2+) ion are tetracyanonickelate(2−) and pentacyanonickelate(3−), both containing nickel in the +2 oxidation state. Similarly, addition of the chloride ion to a solution of hexaaquairon(3+) yields tetrachloroferrate(3−). Both complexes contain iron in the same oxidation state of +3.

Halo Complexes

Probably the most widespread class of complexes involving anionic ligands is that of the complexes of the halide ions—i.e., the fluoride, chloride, bromide, and iodide ions. In addition to forming simple halide salts, such as sodium chloride and nickel difluoride (in which the metal ions are surrounded by halide ions, these in a sense being regarded as coordinated to them), many metals form complex halide salts—such as potassium tetrachloroplatinate(2−), $K_2[PtCl_4]$—that contain discrete complex ions. Most metal ions also form halide complexes in aqueous solution. The stabilities of such complexes span an enormous range—from the alkali-metal ions (lithium, sodium, potassium, and so on), whose formation of halide complexes in aqueous solution can barely be detected, to extremely stable halide complexes, such as the tetraiodomercurate(2−), tetrachlorothallate(1−), and tetrachloropalladate(2−) ions, the extent of whose dissociation is extremely small.

The stabilities of halide complexes reflect a pattern by which metal ions can be divided into two general classes, designated as *A* and *B* or as hard and soft, respectively. (Generally, the electrons in the atoms of the hard elements are considered to form a compact and not easily deformable group, whereas those in the atoms of the soft elements form a looser group—that is, one more easily deformed.) For the former class—which includes Be, Mg, Sc, Cr, Fe, Ni, Cu, In, and Sn—the order of increasing stability of the halide complexes in aqueous solution is iodides < bromides < chlorides < fluorides. Conversely, for the class *B* (or soft) ions—such as Pt, Ag, Cd, Hg, Tl, and Pb—the order of increasing stability of the halide complexes is fluorides < chlorides < bromides < iodides. In contrast to class-*A* metals, those of class *B* also tend to form more stable complexes with sulfur-containing ligands than with oxygen-containing ligands and more stable complexes with phosphorus ligands than with nitrogen ligands.

Carbonyl Complexes

Following the discovery of the first metal carbonyl complex, tetracarbonylnickel, $Ni(CO)_4$, in 1890, many compounds containing carbon monoxide coordinated to transition metals have been prepared and characterized. Such compounds generally contain metal atoms or ions in low oxidation states. The following are some of the more common types of metal carbonyl compounds: (1) simple mononuclear carbonyls of metals in the zero oxidation state, such as tetracarbonylnickel, pentacarbonyliron, and hexacarbonylchromium—highly toxic volatile compounds, the most stable of which have filled valence shells of 18 electrons, (2) salts of anionic and cationic carbonyls, such as tetracarbonylcobaltate(−1) and hexacarbonylmanganese(+1), (3) dinuclear and polynuclear carbonyls, such as bis(tetracarbonylcobalt), the structural formula and (4) mixed complexes containing other ligands in addition to CO: pentacarbonylchloromanganese, tetracarbonylhydridocobalt, and tricarbonylnitrosylcobalt.

Although molecular nitrogen, N_2, is isoelectronic with carbon monoxide (that is, it has the same number and arrangement of electrons), its tendency to form complexes with metals is much smaller. The first complex containing molecular nitrogen as a ligand—i.e., pentaamminenitrogenruthenium(2+), $[Ru(NH_3)_5(N_2)]^{2+}$—was prepared in 1965, and many others have been discovered subsequently. Such complexes have attracted considerable interest because of their possible roles in the chemical and biological fixation of nitrogen.

Nitrosyl Complexes

Nitrosyl complexes can be formed by the reaction of nitric oxide (NO) with many transition metal compounds or by reactions involving species containing nitrogen and oxygen. Some of these complexes have been known for many years—e.g., pentaaquanitrosyliron(2+) ion, $[Fe(H_2O)_5NO]^{2+}$, which formed in the classical brown-ring test for the qualitative detection of nitrate ion; Roussin's red ($K_2[Fe_2S_2(NO)_4]$) and black ($K[Fe_4S_3(NO)_7]$) salts; and sodium pentacyanonitrosylferrate(3−) dihydrate (sodium nitroprusside), $Na_2[Fe(CN)_5NO]·2H_2O$. Such complexes, which can be cationic, neutral, or anionic and which are usually deeply colored (red, brown, purple, or black), have been extensively studied because they pose unique problems of structure and bonding and because they have potential uses as homogeneouscatalysts for a variety of reactions. Because the nitrosonium ion (NO^+) is isoelectronic with carbon monoxide and because its mode of coordination to transition metals is potentially similar to that of carbon monoxide, metal nitrosyls have been recognized as similar to carbonyls and are sometimes formulated as NO^+ complexes. Carbonyl ligands can be replaced by nitric oxide in substitution reactions. Such similarities may be deceptive, however, for the additional electron in neutral nitric oxide requires a more complicated treatment of M-NO bond formation. The NO ligand exhibits several geometries of coordination—linear (e.g., $[IrH(-NO)\{P(C_6H_5)_3\}_3]^+$, $[Mn(CO)_2(NO)\{P(C_6H_5)_3\}_3]$, and $Na_2[Ru(OH)(NO_2)_4(NO)]·2H_2O$); bent (e.g., $[CoNO(NH_3)_5]^{2+}$and $[IrCl_2(NO)\{P(C_6H_5)_3\}_2]$); or both (e.g., $[RuCl(NO)_2\{P(C_6H_5)_3\}_2]^+$). Like CO, NO also can act as a bridging ligand between two (e.g., $[\{Cr(\eta^5{-}C_5H_5)(NO)\}_2(\mu_2{-}NH_2)(\mu_2{-}NO)]$) or three (e.g., $[Mn_3(\eta^5{-}C_5H_5)_3(\mu_2{-}NO)_3(\mu_3{-}NO)]$) metal atoms. (The η^5 indicates that five carbon atoms of the $C_5H_5^-$ group are bonded to the chromium atom.)

Cyano and Isocyano Complexes

Cyano complexes, such as Prussian blue, are among the oldest coordination compounds. In addition to being a pseudohalide, the CN^- ion is isoelectronic with CO, RCN, RNC, N_2, and NO^+ (R is an alkyl group), and metal carbonyls and cyanide complexes are structurally similar. Also, like CO, CN^- enters into π as well as σ bonding with transition metal atoms or ions. Cyano complexes are among the most stable transition metal complexes; the extreme toxicity of CN^- (like that of CO) is due to its irreversible formation of a strong complex with hemoglobin, which prevents oxygen from binding reversibly to hemoglobin, thereby prohibiting the transport and release of oxygen in the body. Similarly, the ability of CN^- to form very stable complexes with silver ($Ag(CN)_2^-$) and gold ($Au(CN)_2^-$) is the basis for its use in the extraction and purification of these metals. As a monodentate ligand, CN^- coordinates (bonds) through carbon as the donor atom, but, as a didentate ligand, it usually coordinates at both ends (C and N) and acts as a bridging ligand (–CN–) to form infinite linear (chain) polymers as in Prussian blue, AgCN, AuCN, $Zn(CN)_2$, and $Cd(CN)_2$.

The cyanide ion forms complexes with transition metals and with zinc, cadmium, and mercury, usually by substitution in aqueous solution with no change in oxidation state. The most important complexes are anionic with the formula $[Mn^+(CN)x]^{(x-n)-}$, where Mn^+ represents a transition metal ion. Examples are $[Ni(CN)_4]^{2-}$, $[Pt(CN)_4]^{2-}$, $[Fe(CN)_6]^{4- \text{ or } 3-}$, $[Co(CN)_6]^{3-}$, $[Pt(CN)_6]^{2-}$, and $[Mo(CN)_8]^{5-, 4-, \text{ or } 3-}$. The free anhydrous parent acids of many of these anions—for example, $H_4[Fe(CN)_6]$ and $H_3[Rh(CN)_6]$—have been isolated.

Cyanide complexes exhibit a variety of coordination numbers and configurations. Metal ions with a d^{10}structure form linear complexes of coordination number 2—as, for example, $[M(CN)_2]^-$(where M = Hg, Ag, or Au)—while the isoelectronic complexes $[Cu(CN)_4]^{3-}$, $[Ag(CN)_4]^{3-}$, $[Zn(CN)_4]^{2-}$, $[Cd(CN)_4]^{2-}$, and $[Hg(CN)_4]^{2-}$ are tetrahedral. All the hexacoordinate complexes are octahedral, while octacoordinate complexes are cubic, dodecahedral, or square antiprismatic. (The dodecahedron and square antiprism are two structures that can be obtained by distorting the simple cube.) For d^2, d^4, d^6, d^8, and d^{10} transition metal ions, the octa-, hepta-, hexa-, penta-, and tetracoordinate complexes, respectively, are species with maximum coordination number.

Mixed complexes of type $[M(CN)_5X]^{n-}$ (where X = H_2O, NH_3, CO, NO, H, or a halogen) also exist. The cyanide ion has the ability to stabilize metal ions in low oxidation states (probably by accepting electron density into its π^*orbitals)—e.g. $[Ni(CN)_4]^{4-}$, which contains nickel in the formal oxidation state of zero. Cyanide complexes have figured prominently in numerous kinetic studies. For example, fast electron-transfer reactions between $[Fe(CN)_6]^{3-}$ and $[Fe(CN)_6]^{4-}$ and between $[Mo(CN)_8]^{3-}$ and $[Mo(CN)_8]^{4-}$ established the outer-sphere mechanism for redox reactions; replacement of water in $[Co(CN)_5H_2O]^{2-}$ established the dissociative mechanism for substitution at a Co(3+) ion.

Transition metals also form complexes with organic cyanides (RCN or ArCN, called nitriles) and organic isocyanides (RNC or ArNC, called isonitriles)—where R and Ar are alkyl and aryl groups, respectively—by reaction of a metal halide, carbonyl, or other complex with the nitrile or isonitrile, respectively. Nitriles and isonitriles appear to be stronger donors of σ electrons than carbon monoxide, but they are capable of extensive back acceptance of π electrons from metals in lower oxidation states—as in $Cr(CNR)_6$ or $Cr(CNAr)_6$ and $Ni(CNR)_4$ or $Ni(CNAr)_4$, which are analogous to the corresponding carbonyls $Cr(CO)_6$ and $Ni(CO)_4$, respectively. Although a bridging isonitrile group has been reported in $(\pi{-}C_5H_5)_2Fe_2(CO)_3(CNC_6H_5)$, this type of bonding is unusual.

Organometallic Complexes

Organometallic complexes are complexes formed between organic groups and metal atoms. They can be divided into two general classes: (1) complexes containing metal-carbon σ bonds and (2) π-bonded metal complexes of unsaturated hydrocarbons—that is, compounds with multiple bonds between carbon atoms.

Isopoly and Heteropoly Anions

The amphoteric metals of groups VB (vanadium, niobium, and tantalum) and VIB (chromium, molybdenum, and tungsten) in the +5 and +6 oxidation states, respectively, form weak acids that readily condense (polymerize) to form anions containing several molecules of the acidanhydride. If these condensed acids contain only one type of acid anhydride, they are called isopoly acids, and their salts are called isopoly salts. The acid anhydrides also can condense with other acids (e.g., phosphoric or silicic acids) to form heteropoly acids, which can form heteropoly salts. The condensation reactions, which occur reversibly in dilute aqueous solution, involve formation of oxo bridges by elimination of water from two molecules of the weak acid. The best-known and simplest example is the condensation of yellow chromate ion (CrO_4^{2-}) to form the orange isopoly dichromate ion ($Cr_2O_7^{2-}$), an equilibrium reaction the extent of which depends on the pH. In acidic solution the isopoly anion $Cr_2O_7^{2-}$, predominates while in basic solution the simple ion CrO_4^{2-} predominates.

$$\underset{\text{basic solution}}{2CrO_4^{2-}+H^+} \rightleftharpoons 2[O_3Cr\text{-}OH]^- \rightleftharpoons \underset{\text{acidic solution}}{[O_3Cr\text{-}O\text{-}CrO_3]^{2-}+H_2O}$$

Heteropoly acids and their salts may be formed by coordination of the central atom with four to six oxo anions, which may be mononuclear (containing one metal ion each), as in $H_7[P(MoO_4)_6]$, or trinuclear (containing three metal ions each), as in $H_3[P(W_3O_{10})_4]$. Incomplete replacement of oxygen atoms in PO_4^{3-} ions by MoO_3 groups can result in dimers (two-molecule polymers), as, for example, $\{OP[O(MoO_3)_3]_3\}_2^{6-}$. About 70 elements can act as central (hetero) atoms in heteropoly anions. Because each element may form more than one heteropoly anion and some of these anions can contain several different heteroatoms, thousands of heteropoly acids exist. Heteroatoms may be primary (these atoms are essential to the polyanion structure and thus not susceptible to chemical exchange) or secondary (these atoms can be removed by chemical reaction from the polyanion structure without destroying it). Heteropoly anions can be regarded as coordination compounds with polyanion ligands; e.g., $[(H_3N)_5Cr(OH_2)]^{3+}$ can be considered the parent of $[(SiW_{11}O_{39})Cr(OH_2)]^{5-}$.

A variety of synthetic procedures are available for the preparation of isopoly acids and salts, which are usually less stable than heteropoly compounds. Heteropolymolybdates and heteropolytungstates are always prepared in solution, usually after acidifying and heating the theoretical amounts of reactants. In general, free heteropoly acids and salts, of which the heteropolymolybdates and heteropolytungstates are the best known, have very high molecular weights (some above 4,000) as compared with other inorganic electrolytes, are very soluble in water and organic solvents, are almost always highly hydrated with several hydrates existing, and are highly colored. Some are strong oxidizing agents that can be reduced to stable, intensely deep blue species (heteropoly blues), which in turn can act as reducing agents, restoring the original color on oxidation. The stoichiometry, oxidation-reduction potentials, and other characteristics of these reactions have been investigated by various methods. The free acids, which are polyprotic (contain several replaceable hydrogen ions), are fairly strong and nearly always stronger than the corresponding acids from which they are derived.

All heteropolymolybdate and heteropolytungstate anions are decomposed in strongly basic solution to form simple molybdate or tungstate ions and either an oxy anion or a hydrous metal oxide of the central metal atom.

$[P_2Mo_{18}O_{62}]^{6-} + 34OH^- \text{----->} 18MoO_4^{2-} + 2HPO_4^{2-} + 16H_2O$

$[NiW_6O_{24}H_6]^{4-} + 8OH^- \text{----->} 6WO_4^{2-} + Ni(OH)_2 + 6H_2O$

Throughout specific ranges of pH and other conditions, most solutions of heteropolymolybdates and heteropolytungstates appear to contain predominantly one distinct species of anion, many of which are remarkably stable and nonlabile.

The first heteropoly compound, $(NH_4)_3[PMo_{12}O_{40}]$, was obtained by the Swedish chemist Jöns Jacob Berzelius in 1826 as a yellow, crystalline precipitate, the formation of which is still used for the classical qualitative detection and quantitative estimation of phosphorus (after conversion to phosphate). By the beginning of the 20th century, hundreds of isopoly and heteropoly compounds were reported, many of which were based on incorrect analyses or failure to detect mixed crystals. Formulas were reported in terms of the old Berzelius dualistic theory as a combination of oxides, such as $3Na_2O{\cdot}Cr_2O_3{\cdot}12MoO_3{\cdot}20H_2O$ for $Na_3CrMo_6O_{24}H_6{\cdot}7H_2O$, and often merely expressed analytical results rather than structure. In addition to their use in analytical chemistry, heteropoly compounds have found use as catalysts, molecular sieves, corrosion inhibitors, photographic fixing agents, and precipitants for basic dyes.

Few structural studies of such compounds were carried out, but this lack did not prevent the elaboration of various unsuccessful theories to account for their structures. In 1907 Werner applied his coordination theory to the structure of 12-tungstosilicic acid, $H_4\,[SiW_{12}O_{40}]$, and its salts by assuming that the central group is an SiO_4^{4-} ion surrounded octahedrally by six $RW_2O_6^+$ groups (R = a unipositive ion), four linked by primary (ionic) and two linked by secondary (coordinate covalent) valences. Difficulties were encountered by this system as well as by the later (1910–21), more elaborate Miolati-Rosenheim theory. Modern conclusive knowledge of the structures of heteropoly compounds did not begin until 1934, with J.F. Keggin's determination of the structure of $H_3\,[PO_4W_{12}O_{36}]{\cdot}5H_2O$ by the most direct means, X-ray diffraction.

The structures of isopoly and heteropoly compounds consist of polyhedrons sharing corners and edges with one another. In heteropolymolybdates or heteropolytungstates, each molybdenum or tungsten atom is located at the centre of an octahedron, each vertex of which is occupied by an oxygen atom. These octahedrons can share corners or edges or both with other MoO_6 or WO_6 octahedrons. In $[Mo_8O_{26}]^{4-}$ eight MoO_6 octahedrons share edges. In $[PMo_{12}O_{40}]^{3-}$ the central phosphorus atom is located at the centre of a PO_4 tetrahedron, which is surrounded by 12 MoO_6 octahedrons, which share corners so that the correct number of oxygen atoms is utilized.

Important Types of Reactions of Coordination Compounds

Acid-base

Coordination to a positive metal ion usually enhances the acidity (i.e., the tendency to lose protons) of hydrogen-containing ligands, such as water and ammonia. Thus, many metal ions in aqueous solution commonly exhibit acidic behavior. Such behavior is exemplified by hydrolysis reactions of the type shown in the following equilibrium:

$$[M(H_2O)_x]^{n+} \rightleftharpoons [M(OH)(H_2O)_{x-1}]^{(n-1)+} + H^+,$$

in which M represents the metal ion, *n* its charge, and *x* the number of coordinated water molecules.

The acidities of such aqua ions depend on the charge, size, and electronic configuration of the metal ion. This dependence is reflected in the values of acid dissociation constants, which range from about 10^{-14} (a value only slightly larger than for pure water, for which the dissociation constant = $10^{-15.7}$) for the hydrated lithium ion, to about 10^{-2} (a value equivalent to that of a fairly strong acid) for the hydrated uranium(4+) ion. Acid-base equilibria are rapidly established in solution, generally within a fraction of a second.

In some cases, hydrolysis of a metal ion may be accompanied by polymerization to form dinuclear or polynuclear hydroxo- or oxygen-bridged complexes.

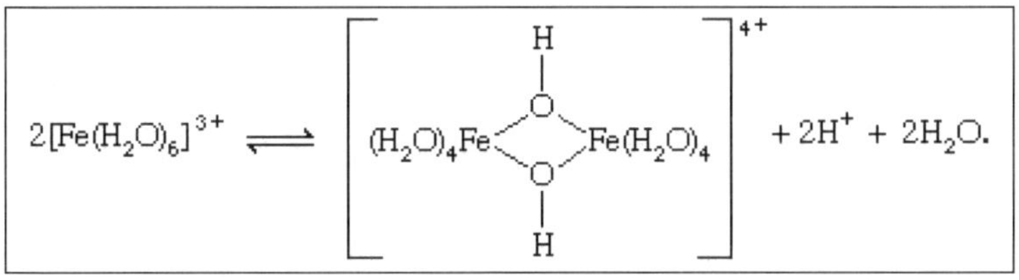

Even very weakly acidic ligands, such as ammonia, can acquire appreciable acidity through coordination to a metal ion. Thus, the hexaammineplatinum(4+) ion dissociates according to the following equilibrium:

$$[Pt(NH_3)_6]^{4+} \rightleftharpoons [Pt(NH_2)(NH_3)_5]^{3+} + H^+.$$

In addition to intrinsic strength, acids and bases have other properties that determine the extent of reactions. According to the hard and soft acids and bases (HSAB) theory, the metal cation and anion are considered to be acids and bases, respectively. Hard acids and bases are small and nonpolarizable, whereas soft acids and bases are larger and more polarizable. Interactions between two hard or soft acids or bases are stronger than ones between one hard and one soft acid or base. The theory can be used to explain solubilities, formation of metallic ores, and some reactions of metal cations with ligands.

Substitution

One of the most general reactions exhibited by coordination compounds is that of substitution, or replacement, of one ligand by another. This process is depicted in a generalized manner by the equation $ML_{x-1}Y + Z \rightarrow ML_{x-1}Z + Y$ for a metal complex of coordination number$x$. The ligands L, Y, and Z may be chemically similar or different. (Charges have been omitted here for simplicity.)

A class of substitution reactions that affords the widest possible comparison of different metal ions is the replacement of water in the coordination spheres of metal-aqua complexes in aqueous solution. The substitution may be by another water molecule (which can be labeled with the isotope oxygen-18 to permit the reaction to be followed) or by a different ligand, such as the chloride ion. Reactions of both types occur as (oxygen-18 is indicated by the symbol $\overset{*}{O}$).

$$[M(H_2O)_x]^{n+} + H_2\overset{*}{O} \longrightarrow [M(H_2O)_{x-1}(H_2\overset{*}{O})]^{n+} + H_2O$$
$$[M(H_2O)_x]^{n+} + Cl^- \longrightarrow [M(H_2O)_{x-1}Cl]^{(n-1)+} + H_2O$$

Many such reactions are extremely fast, and it has been only since 1950, following the development of appropriate experimental methods (including stopped flow, nuclear magnetic resonance, and relaxation spectrometry), that the kinetics and mechanisms of this class of reactions have been extensively investigated. Rates of substitution of metal-aqua ions have been found to span a wide range, the characteristic times required for substitution ranging from less than 10^{-9} second for monopositive ions, such as hydrated potassium ions, to several days for certain more highly charged ions, such as hexaaquachromium(3+) and hexaaquarhodium(3+). The rate of substitution parallels the ease of loss of a water molecule from the coordination sphere of the aqua complex and thus increases with increasing size and with decreasing charge of the metal ion. For transition metal ions, electronic factors also have an important influence on rates of substitution.

There are two limiting mechanisms (or pathways) through which substitution may occur—namely, dissociative and associative mechanisms. In the dissociative mechanism, a ligand is lost from the complex to give an intermediate compound of lower coordination number. This type of reaction path is typical of octahedral complexes, many aqua complexes, and metal carbonyls such as tetracarbonylnickel. An example of a dissociative reaction pathway for an octahedral complex of cobalt is as follows:

$$[Co(CN)_5(H_2O)^2]^{2-} \xrightarrow{-H_2O} [Co(CN)_5]^{2-} \xrightarrow{+I^-} [Co(CN)_5I]^{3-}$$

The associative mechanism for substitution reactions, on the other hand, involves association of an extra ligand with the complex to give an intermediate of higher coordination number; one of the original ligands is then lost to restore the initial coordination number. Substitution reactions of square planar complexes, such as those of the nickel(2+), palladium(2+), and platinum(2+) ions, usually proceed through associative pathways involving intermediates with coordination number five. An example of a reaction following such a pathway is:

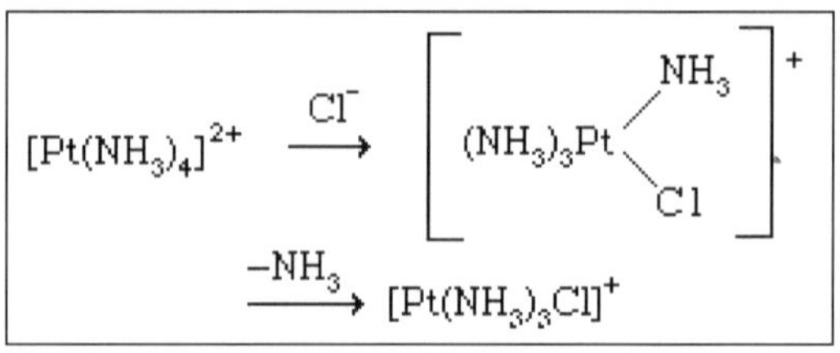

A characteristic feature of this class of reactions is the sensitivity of the rate of substitution of a given ligand to the nature of the ligand in the trans position. The trans ligand activates a ligand for replacement as follows, in decreasing order:

$$CO, CN^-, C_2H_4 > PR_3, H^- > NO_2^-, I^-, SCN^- > Br^-, Cl^- > NH_3, H_2O.$$

The *trans* effect may be used for synthetic purposes; thus, the reaction of the tetrachloroplatinate(2−) ion with ammonia yields *cis*-diamminedichloroplatinum, whereas the reaction of the tetraammineplatinum(2+) ion with the chloride ion gives the *trans* isomer, *trans*-diamminedichloroplatinum. The reactions are:

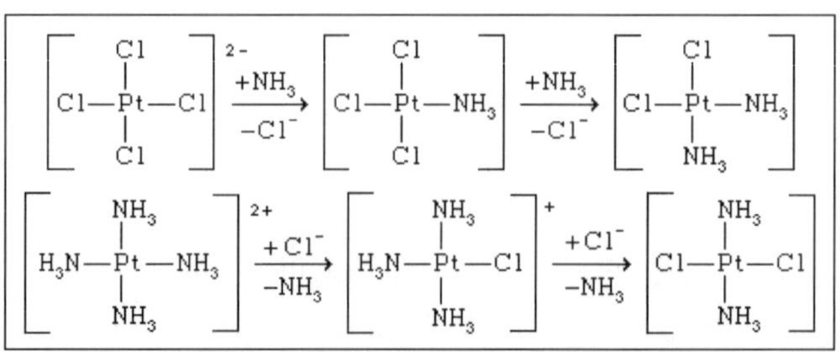

In both reactions, the trans effect causes the introduction of the ligand trans to chloride rather than trans to ammonia.

Lability and Inertness

In considering the mechanisms of substitution (exchange) reactions, Canadian-born American chemist Henry Taube distinguished between complexes that are *labile* (reacting completely in about one minute in 0.1 M solution at room temperature [25 °C, or 77 °F]) and those that are *inert* (under the same conditions, reacting either too slowly to measure or slowly enough to be followed by conventional techniques). These terms refer to kinetics (reaction rates) and should not be confused with the thermodynamic terms *unstable* and *stable*, which refer to equilibrium. For example, as, most cyanide complexes are extremely stable (they possess very small dissociation constants); yet, if their rate of exchange with carbon-14-labeled cyanide, as represented in the following equation:

$$[M(CN)x]^{y-} + x^{14}CN^{-} \rightleftharpoons [M(^{14}CN)x]^{y-} + xCN^{-},$$

is measured, $[Ni(CN)_4]^{2-}$ and $[Hg(CN)_4]^{2-}$ are found to be labile, whereas $[Mn(CN)_6]^{3-}$, $[Fe(CN)_6]^{4-}$, $[Fe(CN)_6]^{3-}$, and $[Cr(CN)_6]^{3-}$ are inert. On the other hand, $[Co(NH_3)_6]^{3+}$, a kinetically inert complex, is thermodynamically stable in acidic solution. Inertness may result from the lack of a suitable low-energy pathway for the reaction. In short, stable complexes possess large positive free energies of reaction (ΔG), whereas inert complexes merely possess large positive free energies of activation (ΔG^{*}).

While the existence of geometric or optical isomers in the solid state or in solution at nonequilibrium concentrations is evidence supporting the inertness of the complex, this does not constitute absolute proof. Conversely, the possibility of intramolecular rearrangement means that failure to isolate geometric isomers or to resolve the racemic mixture into optical isomers is not absolute proof of lability.

Taube has interpreted lability of complexes according to their electronic configuration in terms of VB theory. Labile complexes are either of the outer orbital type (outer *d* orbitals involved in bonding—e.g., sp^3d^2 as opposed to d^2sp^3 [inner orbital] for octahedral complexes) or of the inner orbital type with at least one vacant *d* orbital (available for accommodation of a seventh group during the [associative] substitution reaction).

Isomerization

Coordination compounds that exist in two or more isomeric forms may undergo reactions that convert one isomer to another. Examples are the linkage isomerization and *cis-trans* isomerization reactions depicted below.

$$[Co(ONO)(NH_3)_5]^{2+} \rightarrow [Co(No_2)(NH_3)_5]^{2+}$$

$$\text{cis-}[Co(H_2O)_2(NH_2CH_2CH_2NH_2)_2]^{3+} \rightarrow \textit{trans}[Co(H_2O)_2(NH_2CH_2NH_2)_2]^{3+}$$

The first of these has been shown to proceed intramolecularly (i.e., without dissociation of the nitrite ligand), whereas the second probably occurs through dissociation of one of the water-molecule ligands.

Oxidation-reduction

Transition metals commonly exhibit two or more stable oxidation states, and their complexes accordingly are able to undergo oxidation-reduction reactions. The simplest such reactions involve electron transfer between two complexes, with little if any accompanying rearrangement or chemical change. An example is:

$$[Fe(CN)_6]^{4-}+[IrCl_6]^{2-}\rightarrow[Fe(CN)_6]^{3-}+[IrCl_6]^{3-}.$$

In other cases, oxidation-reduction is accompanied by significant chemical rearrangement. An example is:

$$[CoCl(NH_3)_5]^{2+}+[Cr(H_2O_6]^{2+}+5[H_3O]^{+}\rightarrow[Co(H_2O)_6]^{2+}+[CrCl(H_2O_5)]^{2+}+5[NH_4]^{+}.$$

Two limiting mechanisms of electron transfer, commonly designated outer-sphere and inner-sphere mechanisms, have been recognized. Outer-sphere electron transfer occurs without dissociation or disruption of the coordination sphere of either complex—i.e., through both intact coordination spheres. The first reaction above is of this type. On the other hand, inner-sphere electron transfer—e.g., the second reaction above—proceeds by formation of a dinuclear complex in which the two metal ions are joined by a common bridging ligand (in this case the chloride ion) through which the electron is transferred. Such electron transfer also may occur through polyatomic bridging ligands to which the two metal ions are attached at different sites separated by several atoms, as in the reduction of pentaammine(isonicotinamide)cobalt(3+) by hexaaquachromium(2+) ion through a bridged intermediate:

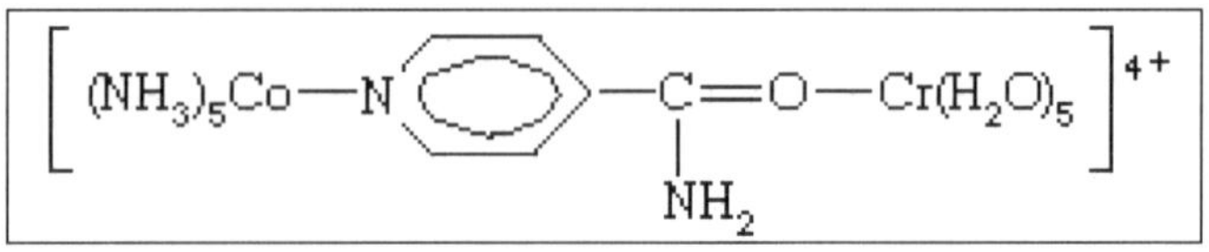

Strikingly large differences in rates of electron transfer are observed even between closely related reactions. Thus, the rate of reduction of the pentaamminebromocobalt(3+) ion by the hexaaquachromium(2+) ion is about 10^7 times higher than that of the acetatopentaamminecobalt(2+) ion by the same chromium ion.

Synthesis of Coordination Compounds

The great variety of coordination compounds is matched by the diversity of methods through which such compounds can be synthesized. Complex halides, for example, may be prepared by direct combination of two halide salts (either in the molten state or in a suitable solvent). Palladium chloride and potassium chloride, for example, react to give the complex potassium tetrachloropalladate(2−), as shown in the following equation:

$$PdCl_2+2KCl\rightarrow K_2[PdCl_4].$$

Another widely used route to coordination compounds is through the direct combination of a metal ion and appropriate ligands in solution. Thus, the addition of a sufficiently high concentration of ammonia to an aqueous solution of a nickel(2+) salt leads, through a series of reactions, to the formation of the hexaamminenickel(2+) ion, which can be precipitated, for example, as the sulfate salt, $[Ni(NH_3)_6]SO_4$.

Complexes of metal ions in high oxidation states are sometimes more readily formed by adding the ligands to a solution of the metal ion in a lower oxidation state in the presence of an oxidizing agent. Thus, addition of ammonia to an aqueous solution of a cobalt(2+) salt in the presence of air or oxygen leads to the formation of cobalt(3+)-ammine complexes such as hexaamminecobalt(3+), $[Co(NH_3)_6]^{3+}$, and pentaammineaquacobalt(3+), $[Co(NH_3)_5(H_2O)]^{3+}$, ions.

Complexes of metals in low oxidation states, such as the carbonyls of metals in their zero oxidation states, can sometimes be prepared by direct combination of the metal with the ligand, as, for example, in the reaction of nickel metal with carbon monoxide.

$$\mathrm{Ni(solid)+4CO \xrightarrow[1\,atm]{25^\circ} Ni(CO)_4}$$

More commonly, a salt of the metal is reduced in the presence of the ligand. An example of this type of synthesis is the reduction of cobalt carbonate with hydrogen in the presence of carbon monoxide to give bis(tetracarbonylcobalt).

$$\mathrm{2CoCO_3+2H_2+8CO \xrightarrow[250\text{-}300\,atm]{120\text{-}200^\circ} Co_2(CO)_8+2CO_2+2H_2O}$$

Similar procedures are applicable to the synthesis of metal sandwich compounds containing cyclopentadienyl and benzene ligands. Dibenzenechromium, for example, can be prepared from chromic chloride, benzene, and aluminum, as shown in the following equation.

$$\mathrm{CrCl_3+2C_6H_6+Al \rightarrow Cr(C_6H_6)_2+AlCl_3}$$

Hydrido complexes of transition metals can be prepared by reactions of suitable precursors either with molecular hydrogen or with suitable reducing agents such as hydrazine or sodium borohydride; for example,

$$\mathrm{2[Co(CN)_5]^{3-}+H_2 \rightarrow 2[CoH(CN)_5]^{3-}}.$$

Transition metal complexes containing metal-carbon bonds can be prepared by a variety of routes, some of the more important of which are illustrated by the following examples.

$$\mathrm{Mn(CO)_5\,Br+LiCH_3 \rightarrow Mn(CO)_5(CH_3)+LiBr},$$

$$\mathrm{[Mn(CO)_5]^- +CH_3Br \rightarrow Mn(CO)_5(CH_3)+Br^-},$$

$$\mathrm{PtHCl[P(CH_3CH_2)_3]_2+CH_2=CH_2 \rightarrow Pt(CH_2CH_3)Cl[P(CH_3CH_2)_3]_2}, \text{and}$$

$$\mathrm{2[Co(CN_5)]^{3-}+CH_3I \rightarrow [Co(CN_5)(CH_3)]^{3-}+[Co(CN)_5I]^{3-}}$$

Denticity

Denticity refers to the number of donor groups in a single ligand that bind to a central atom in a coordination complex. In many cases, only one atom in the ligand binds to the metal, so the denticity equals one, and the ligand is said to be monodentate (sometimes called unidentate). Ligands with more than one bonded atom are called polydentate or multidentate. The word *denticity* is derived from *dentis*, the Latin word for tooth. The ligand is thought of as biting the metal at one or more linkage points. The denticity of a ligand is described with the Greek letter κ ('kappa'). For example, κ^6-EDTA describes an EDTA ligand that coordinates through 6 non-contiguous atoms.

Denticity is different from hapticity because hapticity refers exclusively to ligands where the coordinating atoms are contiguous. In these cases the η ('eta') notation is used. Bridging ligands use the μ ('mu') notation.

Classes

Polydentate ligands are chelating agents and classified by their denticity. Some atoms cannot form the maximum possible number of bonds a ligand could make. In that case one or more binding sites of the ligand are unused. Such sites can be used to form a bond with another chemical species.

- Bidentate (also called didentate) ligands bind with two atoms, an example being ethylenediamine.

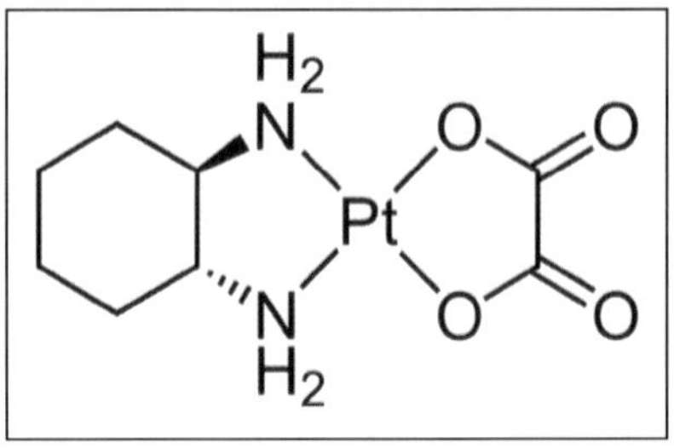

Structure of the pharmaceutical Oxaliplatin, which features two different bidentate ligands.

- Tridentate ligands bind with three atoms, an example being terpyridine. Tridentate ligands usually bind via two kinds of connectivity, called "mer" and "fac." "fac" stands for facial, the donor atoms are arranged on a triangle around one face of the octahedron. "mer" stands for meridian, where the donor atoms are stretched out around one half of the octahedron. Cyclic tridentate ligands such as TACN and 9-ane-S3 bind in a facial manner.

- Tetradentate ligands bind with four donor atoms, an example being triethylenetetramine (abbreviated trien). For different central metal geometries there can be different numbers of isomers depending on the ligand's topology and the geometry of the metal center. For octahedral metals, the linear tetradentate trien can bind via three geometries. Tripodal tetradentate ligands, e.g. tris(2-aminoethyl)amine, are more constrained, and on octahedra leave two cis sites (adjacent to each other). Many naturally occurring macrocyclic ligands are tetradentative, an example being the porphyrin in heme. On an octahedral metal these leave two vacant sites opposite each other.

- Pentadentate ligands bind with five atoms, an example being ethylenediaminetriacetic acid.
- Hexadentate ligands bind with six atoms, an example being EDTA (although it can bind in a tetradentate manner).
- Ligands of denticity greater than 6 are well known. The ligands 1,4,7,10-tetraazacyclododecane-1,4,7,10-tetraacetate (DOTA) and diethylene triamine pentaacetate (DTPA) are octadentate. They are particularly useful for binding lanthanide ions, which typically have coordination numbers greater than 6.

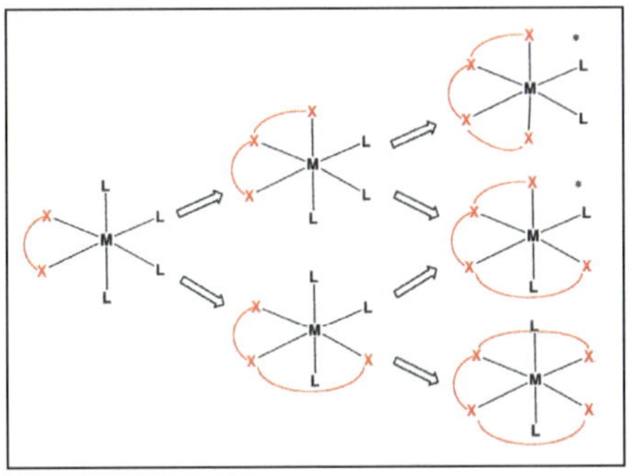

Relationship between "linear" bi-, tri- and tetradentate ligands (red) bound to an octahedral metal center. The structures marked with * are chiral owing to the backbone of the tetradentate ligand.

Stability Constants

In general, the stability of a metal complex correlates with the denticity of the ligands, which can be attributed to the chelate effect. Polydentate ligands such as hexa- or octadentate ligands tend to bind metal ions more strongly than ligands of lower denticity, primarily due to entropic factors. Stability constants are a quantitative measure to assess the thermodynamic stability of coordination complexes.

Hapticity

Hapticity is the coordination of a ligand to a metal center via an uninterrupted and contiguous series of atoms. The hapticity of a ligand is described with the Greek letter η ('eta'). For example, η^2 describes a ligand that coordinates through 2 contiguous atoms. In general the η-notation only applies when multiple atoms are coordinated (otherwise the κ-notation is used). In addition, if the ligand coordinates through multiple atoms that are not contiguous then this is considered denticity (not hapticity), and the κ-notation is used once again. When naming complexes care should be taken not to confuse η with μ ('mu'), which relates to bridging ligands.

Examples:

The η-notation is encountered in many co-ordination compounds.

- Side-on bonding of molecules containing σ-bonds like H_2:
  - $W(CO)_3(P^iPr_3)_2(\eta^2\text{-}H_2)$.

- Side-on bonded ligands containing multiple bonded atoms, e.g. ethylene in Zeise's salt or with fullerene, which is bonded through donation of the π-bonding electrons:
  - $K[PtCl_3(\eta^2\text{-}C_2H_4)]{\cdot}H_2O$.
- Related complexes containing bridging π-ligands:
  - $(\mu\text{-}\eta^2{:}\eta^2\text{-}C_2H_2)Co_2(CO)_6$ and $(Cp^*{}_2Sm)_2(\mu\text{-}\eta^2{:}\eta^2\text{-}N_2)$.
  - Dioxygen in bis{(trispyrazolylborato)copper(II)}$(\mu\text{-}\eta^2{:}\eta^2\text{-}O_2)$.

  Note that with some bridging ligands, an alternative bridging mode is observed, e.g. κ^1,κ^1, like in $(Me_3SiCH_2)_3V(\mu\text{-}N_2\text{-}\kappa^1(N),\kappa^1(N'))V(CH_2SiMe_3)_3$ contains a bridging dinitrogen molecule, where the molecule is end-on coordinated to the two metal centers.
- The bonding of π-bonded species can be extended over several atoms, e.g. in allyl, butadiene ligands, but also in cyclopentadienyl or benzene rings can share their electrons.
- Apparent violations of the 18-electron rule sometimes are explicable in compounds with unusual hapticities:
  - The 18-VE complex $(\eta^5\text{-}C_5H_5)Fe(\eta^1\text{-}C_5H_5)(CO)_2$ contains one η^5 bonded cyclopentadienyl, and one η^1 bonded cyclopentadienyl.
  - Reduction of the 18-VE compound $[Ru(\eta^6\text{-}C_6Me_6)_2]^{2+}$ (where both aromatic rings are bonded in an η^6-coordination), results in another 18VE compound: $[Ru(\eta^6\text{-}C_6Me_6)(\eta^4\text{-}C_6Me_6)]$.
- Examples of polyhapto coordinated heterocyclic and inorganic rings: $Cr(\eta^5\text{-}C_4H_4S)(CO)_3$ contains the sulfur heterocycle thiophene and $Cr(\eta^6\text{-}B_3N_3Me_6)(CO)_3$ contains a coordinated inorganic ring (B_3N_3 ring).

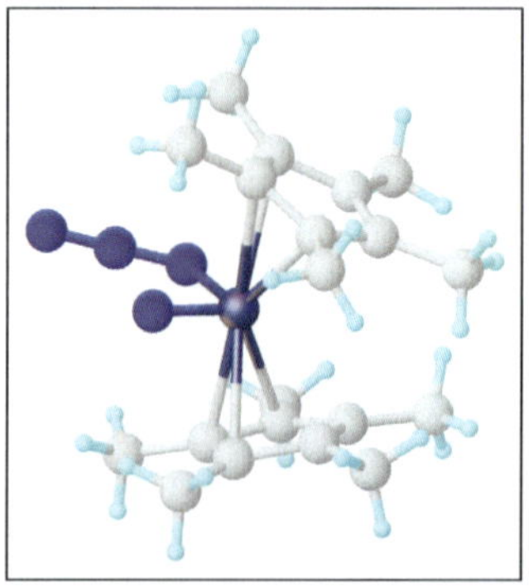

Structure of $(\eta^3\text{-}C_5Me_5)_2Mo(N)(N_3)$.

Electrons Donated by "π- Ligands" vs. Hapticity

Ligand	Electrons contributed (neutral counting)	Electrons contributed (ionic counting)
η^1 Allyl	1	2
η^3-Allyl cyclopropenyl	3	4

η²-Butadiene	2	2
η⁴-Butadiene	4	4
η¹-cyclopentadienyl	1	2
η³-cyclopentadienyl	3	4
η⁵-cyclopentadienyl pentadienyl cyclohexadienyl	5	6
η²-Benzene	2	2
η⁴-Benzene	4	4
η⁶-Benzene	6	6
η⁷-Cycloheptatrienyl	7	6 or 10
η⁸-Cyclooctatetraenyl	8	10

Changes in Hapticity

The hapticity of a ligand can change in the course of a reaction.

E.g. in a redox reaction:

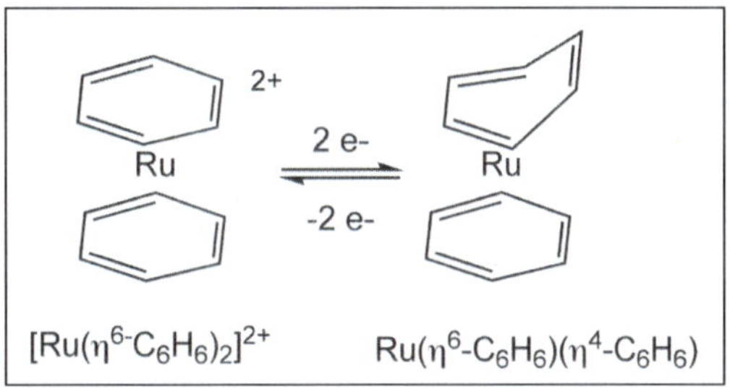

Here one of the η⁶-benzene rings changes to a η⁴-benzene.

Similarly hapticity can change during a substitution reaction:

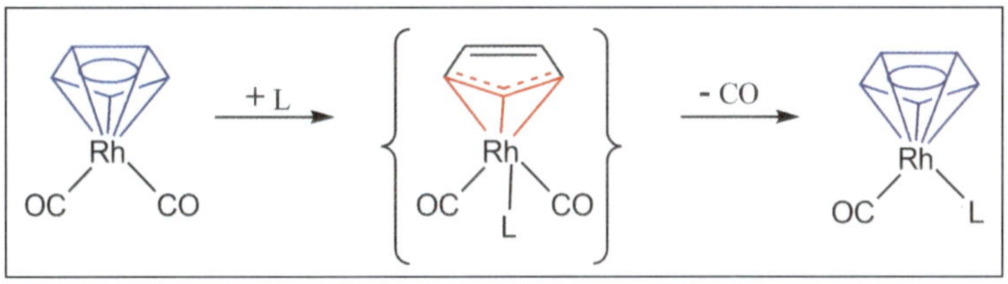

Here the η⁵-cyclopentadienyl changes to an η³-cyclopentadienyl, giving room on the metal for an extra 2-electron donating ligand 'L'. Removal of one molecule of CO and again donation of two more electrons by the cyclopentadienyl ligand restores the η⁵-cyclopentadienyl. The so-called indenyl effect also describes changes in hapticity in a substitution reaction.

Hapticity vs. Denticity

Hapticity must be distinguished from denticity. Polydentate ligands coordinate via multiple coordination sites within the ligand. In this case the coordinating atoms are identified using the κ-notation, as for example seen in coordination of 1,2-bis(diphenylphosphino)ethane ($Ph_2PCH_2CH_2PPh_2$), to $NiCl_2$ as dichloro[ethane-1,2-diylbis(diphenylphosphane)-κ²P]nickel(II). If the coordinating atoms are contiguous (connected to each other), the η-notation is used, as e.g. in titanocene dichloride: dichlorobis(η^5-2,4-cyclopentadien-1-yl)titanium.

Hapticity and Fluxionality

Molecules with polyhapto ligands are often fluxional, also known as stereochemically non-rigid. Two classes of fluxionality are prevalent for organometallic complexes of polyhapto ligands:

- Case 1, typically: When the hapticity value is less than the number of sp^2 carbon atoms. In such situations, the metal will often migrate from carbon to carbon, maintaining the same net hapticity. The η^1-C_5H_5 ligand in (η^5-C_5H_5)Fe(η^1-C_5H_5)$(CO)_2$ rearranges rapidly in solution such that Fe binds alternatingly to each carbon atom in the η^1-C_5H_5 ligand. This reaction is degenerate and, in the jargon of organic chemistry, it is an example of a sigmatropic rearrangement. A related example is Bis(cyclooctatetraene)iron, in which the η^4- and η^6-C_8H_8 rings interconvert.
- Case 2, typically: Complexes containing cyclic polyhapto ligands with maximized hapticity. Such ligands tend to rotate. A famous example is ferrocene, Fe(η^5-C_5H_5)$_2$, wherein the Cp rings rotate with a low energy barrier about the principal axis of the molecule that "skewers" each ring. This "ring whizzing" explains, inter alia, why only one isomer can be isolated for Fe(η^5-C_5H_4Br)$_2$. In this case, the rotamers are not necessarily degenerate, but the rotational barriers have low energies of activation.

Chelate

Chelate is any of a class of coordination or complex compounds consisting of a central metal atom attached to a large molecule, called a ligand, in a cyclic or ring structure. An example of a chelate ring occurs in the ethylenediamine-cadmium complex:

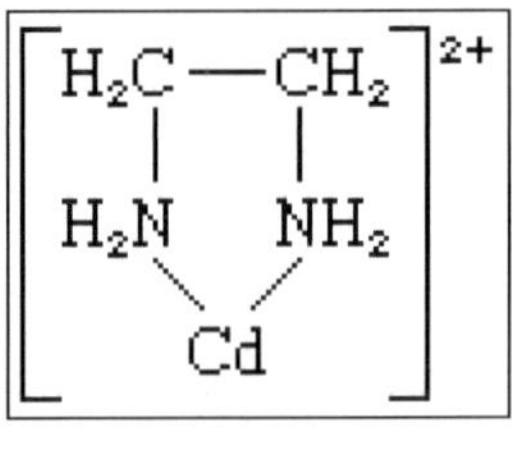

The ethylenediamine ligand has two points of attachment to the cadmium ion, thus forming a ring; it is known as a didentate ligand. (Three ethylenediamine ligands can attach to the Cd^{2+} ion, each one forming a ring as depicted above.) Ligands that can attach to the same metal ion at two or more points are known as polydentate ligands. All polydentate ligands are chelating agents.

Chelates are more stable than nonchelated compounds of comparable composition, and the more extensive the chelation—that is, the larger the number of ring closures to a metal atom—the more stable the compound. This phenomenon is called the chelate effect; it is generally attributed to an increase in the thermodynamic quantity called entropy that accompanies chelation. The stability of a chelate is also related to the number of atoms in the chelate ring. In general, chelates containing five- or six-membered rings are more stable than chelates with four-, seven-, or eight-membered rings.

In medical practice, chelating agents, particularly salts of EDTA, or edetic (ethylenediaminetetraacetic) acid, are widely used for direct treatment of metal poisoning because they bind the toxic metal ions more strongly than do the vulnerable components of the living organism. Chelating agents are also employed as extractants in industrial and laboratory separation of metals and as metal-ion buffers and indicators in analytical chemistry. Many commercial dyes and a number of biological substances, including chlorophyll and hemoglobin, are chelate compounds.

Chelation

Chelation is a type of bonding of ions and molecules to metal ions. It involves the formation or presence of two or more separate coordinate bonds between a polydentate (multiple bonded) ligand and a single central atom. These ligands are called chelants, chelators, chelating agents, or sequestering agents. They are usually organic compounds.

Chelation is useful in applications such as providing nutritional supplements, in chelation therapy to remove toxic metals from the body, as contrast agents in MRI scanning, in manufacturing using homogeneous catalysts, in chemical water treatment to assist in the removal of metals, and in fertilizers.

Chelate Effect

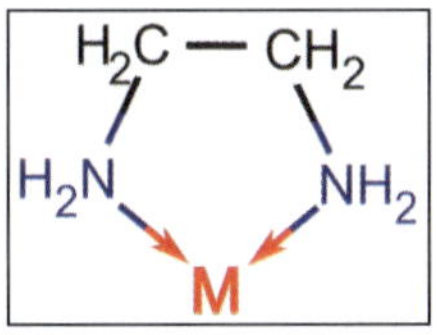

Ethylenediamine ligand chelating to a metal with two bonds.

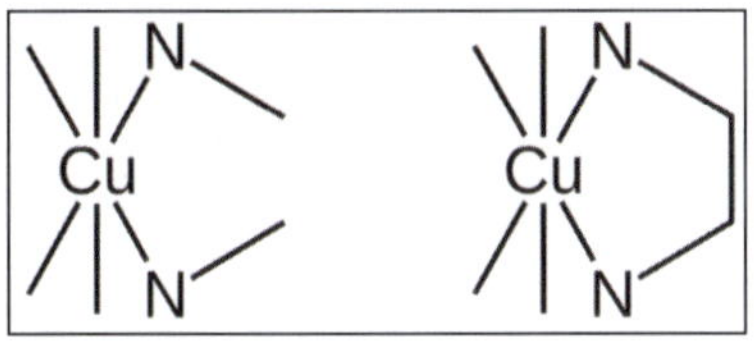

Cu^{2+} complexes with nonchelating methylamine (left) *and* chelating ethylenediamine (right) ligands.

The chelate effect is the enhanced affinity of chelating ligands for a metal ion compared to the affinity of a collection of similar nonchelating (monodentate) ligands for the same metal.

The thermodynamic principles underpinning the chelate effect are illustrated by the contrasting affinities of copper (II) for ethylenediamine (en) vs. methylamine.

$$Cu^{2+} + en \rightleftharpoons [Cu(en)]^{2+}$$

$$Cu^{2+} + 2MeNH_2 \rightleftharpoons [Cu(MeNH_2)_2]^{2+}$$

In ($Cu^{2+} + en \rightleftharpoons [Cu(en)]^{2+}$) the ethylenediamine forms a chelate complex with the copper ion. Chelation results in the formation of a five-membered CuC_2N_2 ring. In ($Cu^{2+} + 2MeNH_2 \rightleftharpoons [Cu(MeNH_2)_2]^{2+}$)

the bidentate ligand is replaced by two monodentate methylamine ligands of approximately the same donor power, indicating that the Cu–N bonds are approximately the same in the two reactions.

The thermodynamic approach to describing the chelate effect considers the equilibrium constant for the reaction: The larger the equilibrium constant, the higher the concentration of the complex.

$$[Cu(en)] = \beta_{11}[Cu][en]$$

$$[Cu(MeNH_2)_2] = \beta_{12}[Cu][MeNH_2]^2$$

Electrical charges have been omitted for simplicity of notation. The square brackets indicate concentration, and the subscripts to the stability constants, β, indicate the stoichiometry of the complex. When the analytical concentration of methylamine is twice that of ethylenediamine and the concentration of copper is the same in both reactions, the concentration [Cu(en)] is much higher than the concentration $[Cu(MeNH_2)_2]$ because $\beta_{11} \gg \beta_{12}$.

An equilibrium constant, *K*, is related to the standard Gibbs free energy, $\Delta G^\ominus$ by:

$$\Delta G^\ominus = -RT \ln K = it\Delta H^\ominus - T\Delta S^\ominus$$

where *R* is the gas constant and *T* is the temperature in kelvins. $\Delta H^\ominus$ is the standard enthalpy change of the reaction and $\Delta S^\ominus$ is the standard entropy change.

Since the enthalpy should be approximately the same for the two reactions, the difference between the two stability constants is due to the effects of entropy. In equation above there are two particles on the left and one on the right, whereas in equation $Cu^{2+} + en \rightleftharpoons [Cu(en)]^{2+}$ and $Cu^{2+} + 2MeNH_2 \rightleftharpoons [Cu(MeNH_2)_2]^{2+}$ there are three particles on the left and one on the right. This difference means that less entropy of disorder is lost when the chelate complex is formed with bidentate ligand than when the complex with monodentate ligands is formed. This is one of the factors contributing to the entropy difference. Other factors include solvation changes and ring formation. Some experimental data to illustrate the effect are shown in the following table.

Equilibrium	log β	$\Delta G^\ominus$	$\Delta H^\ominus$/kJ mol^{-1}	$-T\Delta S^\ominus$/kJ mol^{-1}
$Cu^{2+} + 2\ MeNH_2 \rightleftharpoons Cu(MeNH_2)_2^{\ 2+}$	6.55	−37.4	−57.3	19.9
$Cu^{2+} + en \rightleftharpoons Cu(en)^{2+}$	10.62	−60.67	−56.48	−4.19

These data confirm that the enthalpy changes are approximately equal for the two reactions and that the main reason for the greater stability of the chelate complex is the entropy term, which is much less unfavorable. In general it is difficult to account precisely for thermodynamic values in terms of changes in solution at the molecular level, but it is clear that the chelate effect is predominantly an effect of entropy.

In Nature

Numerous biomolecules exhibit the ability to dissolve certain metal cations. Thus, proteins, polysaccharides, and polynucleic acids are excellent polydentate ligands for many metal ions. Organic compounds such as the amino acids glutamic acid and histidine, organic diacids such as malate, and polypeptides such as phytochelatin are also typical chelators. In addition to these adventitious chelators, several biomolecules are specifically produced to bind certain metals.

In Biochemistry and Microbiology

Virtually all metalloenzymes feature metals that are chelated, usually to peptides or cofactors and prosthetic groups. Such chelating agents include the porphyrin rings in hemoglobin and chlorophyll. Many microbial species produce water-soluble pigments that serve as chelating agents, termed siderophores. For example, species of *Pseudomonas* are known to secrete pyochelin and pyoverdine that bind iron. Enterobactin, produced by *E. coli*, is the strongest chelating agent known. The marine mussels use metal chelation esp. Fe^{3+} chelation with the Dopa residues in mussel foot protein-1 to improve the strength of the threads that it uses to secure itself to surfaces.

In Geology

In earth science, chemical weathering is attributed to organic chelating agents (*e.g.*, peptides and sugars) that extract metal ions from minerals and rocks. Most metal complexes in the environment and in nature are bound in some form of chelate ring (*e.g.*, with a humic acid or a protein). Thus, metal chelates are relevant to the mobilization of metals in the soil, the uptake and the accumulation of metals into plants and microorganisms. Selective chelation of heavy metals is relevant to bioremediation (*e.g.*, removal of ^{137}Cs from radioactive waste).

Medical Applications

Nutritional Supplements

In the 1960s, scientists developed the concept of chelating a metal ion prior to feeding the element to the animal. They believed that this would create a neutral compound, protecting the mineral from being complexed with insoluble salts within the stomach, which would render the metal unavailable for absorption. Amino acids, being effective metal binders, were chosen as the prospective ligands, and research was conducted on the metal–amino acid combinations. The research supported that the metal–amino acid chelates were able to enhance mineral absorption.

During this period, synthetic chelates such as ethylenediaminetetraacetic acid (EDTA) were being developed. These applied the same concept of chelation and did create chelated compounds; but these synthetics were too stable and not nutritionally viable. If the mineral was taken from the EDTA ligand, the ligand could not be used by the body and would be expelled. During the expulsion process the EDTA ligand randomly chelated and stripped another mineral from the body.

According to the Association of American Feed Control Officials (AAFCO), a metal–amino acid chelate is defined as the product resulting from the reaction of metal ions from a soluble metal salt with amino acids, with a mole ratio in the range of 1–3 (preferably 2) moles of amino acids for one mole of metal. The average weight of the hydrolyzed amino acids must be approximately 150 and the resulting molecular weight of the chelate must not exceed 800 Da.

Since the early development of these compounds, much more research has been conducted, and has been applied to human nutrition products in a similar manner to the animal nutrition experiments that pioneered the technology. Ferrous bis-glycinate is an example of one of these compounds that has been developed for human nutrition.

Dental and Oral Application

First-generation dentin adhesives were first designed and produced in the 1950s. These systems were based on a co-monomer chelate with calcium on the surface of the tooth and generated very weak water resistance chemical bonding (2–3 MPa).

Heavy-metal Detoxification

Chelation therapy is an antidote for poisoning by mercury, arsenic, and lead. Chelating agents convert these metal ions into a chemically and biochemically inert form that can be excreted. Chelation using calcium disodium EDTA has been approved by the U.S. Food and Drug Administration (FDA) for serious cases of lead poisoning. It is not approved for treating "heavy metal toxicity".

Although beneficial in cases of serious lead poisoning, use of disodium EDTA (edetate disodium) instead of calcium disodium EDTA has resulted in fatalities due to hypocalcemia. Disodium EDTA is not approved by the FDA for any use, and all FDA-approved chelation therapy products require a prescription.

Pharmaceuticals

Chelate complexes of gadolinium are often used as contrast agents in MRI scans, although iron particle and manganese chelate complexes have also been explored. Bifunctional chelate complexes of zirconium, gallium, fluorine, copper, yttrium, bromine, or iodine are often used for conjugation to monoclonal antibodies for use in antibody-based PET imaging. These chelate complexes often employ the usage of hexadentate ligands such as desferrioxamine B (DFO), according to Meijs *et al.*, and the gadolinium complexes often employ the usage of octadentate ligands such as DTPA, according to Desreux *et al.* Auranofin, a chelate complex of gold, is used in the treatment of rheumatoid arthritis, and penicillamine, which forms chelate complexes of copper, is used in the treatment of Wilson's disease and cystinuria, as well as refractory rheumatoid arthritis.

Other Medical Applications

Chelation in the intestinal tract is a cause of numerous interactions between drugs and metal ions (also known as "minerals" in nutrition). As examples, antibiotic drugs of the tetracycline and quinolone families are chelators of Fe^{2+}, Ca^{2+}, and Mg^{2+} ions.

EDTA, which binds to calcium, is used to alleviate the hypercalcimia that often results from band keratopathy. The calcium may then be removed from the cornea, allowing for some increase in clarity of vision for the patient.

Industrial and Agricultural Applications

Catalysis

Homogeneous catalysts are often chelated complexes. A representative example is the use of BINAP (a bidentate phosphine) in Noyori asymmetric hydrogenation and asymmetric isomerization. The latter has the practical use of manufacture of synthetic (–)-menthol.

Water Softening

Citric acid is used to soften water in soaps and laundry detergents. A common synthetic chelator is EDTA. Phosphonates are also well-known chelating agents. Chelators are used in water treatment programs and specifically in steam engineering, *e.g.*, boiler water treatment system: *Chelant Water Treatment system*. Although the treatment is often referred to as "softening," chelation has little effect on the water's mineral content, other than to make it soluble. What does change is the water's pH level, which is lowered.

Fertilizers

Metal chelate compounds are common components of fertilizers to provide micronutrients. These micronutrients (manganese, iron, zinc, copper) are required for the health of the plants. Most fertilizers contain phosphate salts that, in the absence of chelating agents, typically convert these metal ions into insoluble solids that are of no nutritional value to the plants. EDTA is the typical chelating agent that keeps these metal ions in a soluble form.

Chirality

Stereoisomers are isomers that differ in spatial arrangement of atoms, rather than order of atomic connectivity. One of their most interesting type of isomer is the mirror-image stereoisomers, a non-superimposable set of two molecules that are mirror image of one another. The existance of these molecules are determined by concept known as chirality. The word "chiral" was derived from the Greek word for hand, because our hands display a good example of chirality since they are non-superimposable mirror images of each other.

The opposite of chiral is achiral. Achiral objects are superimposable with their mirror images. For example, two pieces of paper are achiral. In contrast, chiral molecules, like our hands, are non superimposable mirror images of each other.

Try to line up your left hand perfectly with your right hand, so that the palms are both facing in the same directions. Spend about a minute doing this. The same thing applies to some molecules

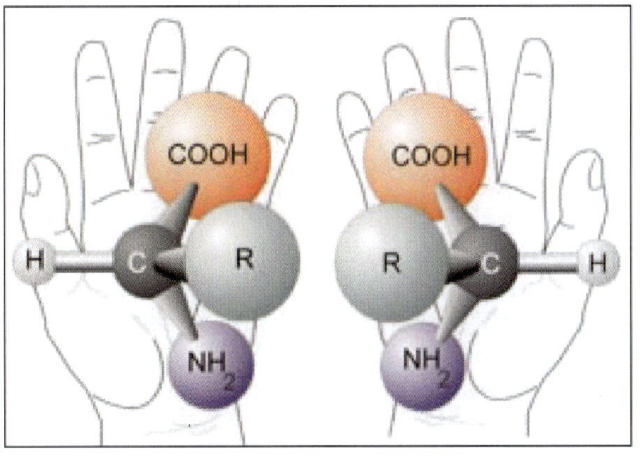

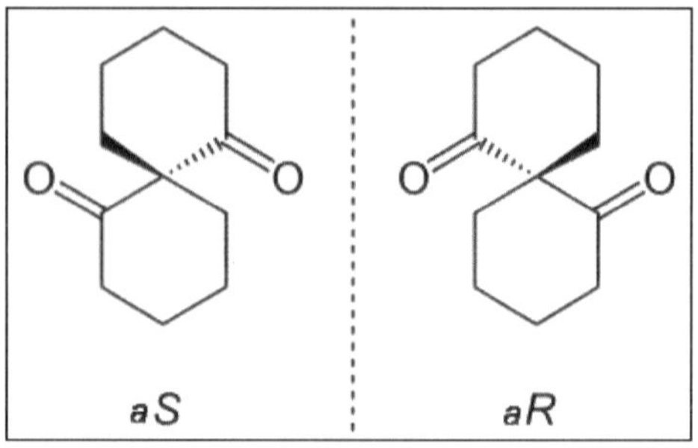

A Chiral molecule has a mirror image that cannot line up with it perfectly- the mirror images are non superimposable. The mirror images are called enantiomers.

But why are chiral molecules so interesting? A chiral molecule and its enantiomer have the same chemical and physical properties (boiling point, melting point, polarity, density etc.) It turns out that many of our biological molecules such as our DNA, amino acids and sugars, are chiral molecules.

It is pretty interesting that our hands seem to serve the same purpose but most people are only able to use one of their hands to write. Similarily this is true with chiral biological molecules and interactions. Just like your left hand will not fit properly in your right glove, one of the enantiomers of a molecule may not work the same way in your body.

This must mean that enantiomers have properties that make them unique to their mirror images. One of these properties is that they cannot have a plane of symmetry or an internal mirror plane. So, a chiral molecule cannot be divided in two mirror image halves. Another property of chiral molecules is optical activity.

Optical Activity

Chiral molecules are very similar to each other since they have the same components to them. The only thing that obviously differs is their arrangement in space. As a result of this similarity, it is very hard to distinguish chiral molecules from each other when we try to compare their properties such as boiling points, melting points and densities.

However, we can differentiate them by their optical activity. When a plane-polarized light is passed through one of the 2 enantiomers of a chiral molecule that molecule rotates light in a certain direction. When the same plane polarized light is passed through the other enantiomer, that enantionmer rotates light by the same amount but in the opposite direction.

If one enantiomer rotates the light counterclockwise, the other would rotate it clockwise. Because chiral molecules are able to rotate the plane of polarization differently by interacting with the electric field differently, they are said to be optically active. In general molecules that rotate light in differen directions are called optical isomers.

- Dextrorotatory (+ enantiomer) – rotates the plane polarized light clockwise (when viewing towards the light source).
- Levorotatory (- enantiomer) – rotates the plane polarized light counterclockwise.

Circular Dichroism

Another property of chiral molecules is called circular dichroism (CD). This pertains to their differential absorption of left and right circularly polarized light. When left and right circularly polarized light passes through chiral molecules, the absorption coefficients differ so that the change in absorption coefficients does not equal zero.

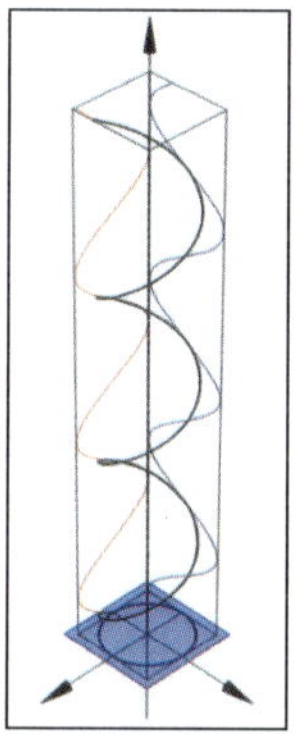

$$\Delta A = A_L - A_R$$

Where ΔA is the difference between absorbance of left circularly polarized (LCP) and right circularly polarized (RCP) light (this is what is usually measured). Where [J] = molar concentration of the sample and l is the path length.

The CD signals of chiral molecules can give important information and this information can be used for visible and ultraviolet spectroscopy. Every chiral molecule shows a particular CD spectrum. By looking at the distinctive spectra of molecules such as proteins and DNA, we can obtain useful information about their secondary structures and see how they differ. An example of a CD spectrum comparing the differences between A-DNA and B-DNA.

Ligand Substitution Reaction

Ligand substitution is the first reaction one typically encounters in an organometallic chemistry. In general, ligand substitution involves the exchange of one ligand for another, with no change in oxidation state at the metal center. The incoming and outgoing ligands may be L- or X-type, but the charge of the complex changes if the ligand type changes. Keep charge conservation in mind when writing out ligand substitutions.

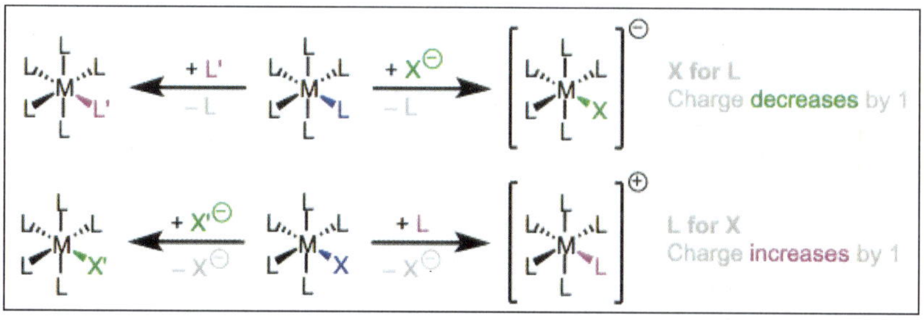

Charge is conserved in ligand substitution reactions. Four general types of substitution are shown here—note the X-for-L and L-for-X types.

How do we know when a ligand substitution reaction is favorable? The thermodynamics of the reaction depend on the relative strength of the two metal-ligand bonds, and the stability of the departing and incoming ligands (or salt sof the ligand, if they're X type). It's often useful to think of X-for-X substitutions like acid-base reactions, with the metal and spectator ligands serving as a "glorified proton." Like acid-base equilibria in organic chemistry, we look to the relative stability of the two charged species (the free ligands) to draw conclusions. Of course we don't necessarily need to rely just on primal thermodynamics to drive ligand substitution reactions. Photochemistry, neighboring-group participation, and other tools can facilitate otherwise difficult substitutions.

Ligand substitution is characterized by a continuum of mechanisms bound by associative (A) and dissociative (D) extremes. At the associative extreme, the incoming ligand first forms a bond to the metal, then departing ligand takes its lone pair and leaves. At the dissociative extreme, the order of events is opposite—the departing ligand leaves, then the incoming ligand comes in. Associative substitution is common for 16-electron complexes (like d8 complexes of Ni, Pd, and Pt), while dissociative substitution is the norm for 18-electron complexes. Then again, reality is often more complicated than these extremes. In some cases, evidence is available for simultaneous dissociation and association, and this mechanism has been given the name interchange (IA or ID).

We'll explore ligand substitution reactions and mechanisms in detail. We'd like to be able to (a) predict whether a mechanism is likely to be associative or dissociative; (b) propose a reasonable mechanism from given experimental data; and (c) describe the results we'd expect given a particular mechanism.

Mechanism of Ligand Substitution

There are two basic steps in ligand substitution: association and dissociation. Association, in this case, refers to the binding of a ligand to the metal. The ligand donates an electron pair to the metal and the two molecules come together to form a new bond. Dissociation, in this case, refers to the release of a ligand from a metal. The metal-ligand bond breaks and the ligand leaves with its electron pair.

Two mechanistic possibilities seem pretty obvious. Either the new ligand binds first and then the old one leaves, or the old ligand leaves first and then the new one binds.

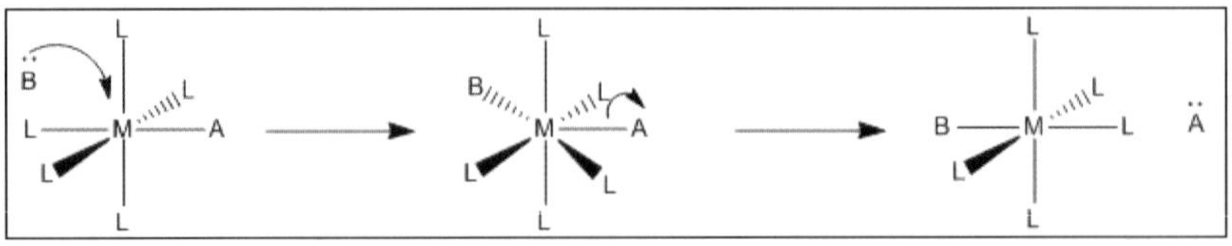

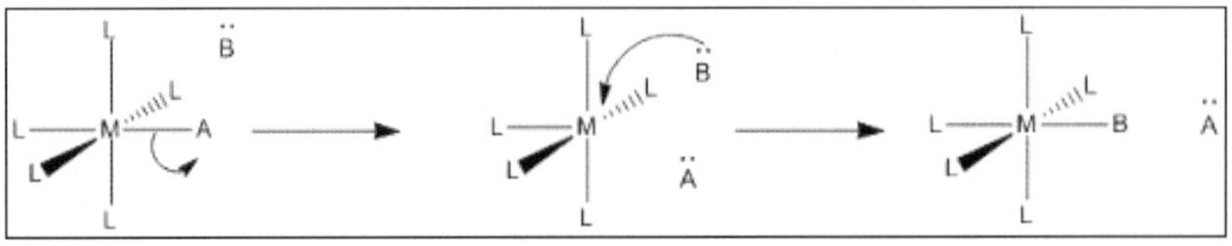

- Associative mechanism is association first. The new ligand binds and then the old one leaves.
- Dissociative mechanism is dissociation first. The old ligand leaves and then the new one binds.

Knowing the mechanism is important because the mechanism has an impact on what factors affect the reaction. For example, if the reaction is associative, adding lots more new ligand may speed up the reaction, because then it becomes more likely that the new ligand will find the metal complex and bind with it. However, if the old ligand is supposed to leave before the new ligand arrives, then it doesn't matter how much new ligand is around. It has to wait for the old ligand to leave before it can bind, anyway, so adding a lot more new ligand won't speed things up.

Photoredox Processes in Coordination Compounds

Electron transfer reactions of coordination compounds have been the subject of intense investigations by inorganic chemists both experimentally and theoretically for many years. Much of the earlier studies pioneered by Taube and coworker's are concerned with the thermal electron transfer reactions and the theoretical framework was worked out by Marcus, Hush and others. In recent years considerable amount of work has been done on excited state electron transfer processes of coordination compounds. Several attempts have been made to extend these studies to the photo-electrochemistry of liquid junction semiconductors. It looks promising that photo-induced redox reactions should serve to convert solar energy into fuels and electricity.

Excited States of Coordination Compounds

Reasonably good theoretical models exist to understand electronic excited states of coordination compounds where the excitation is localised to metal centered orbitals and to derive the nature of the energy levels of the compounds. The electronic excited states of a typical octahedral coordination compound are schematically depicted in figure and charge-transfer or electron transfer bands.

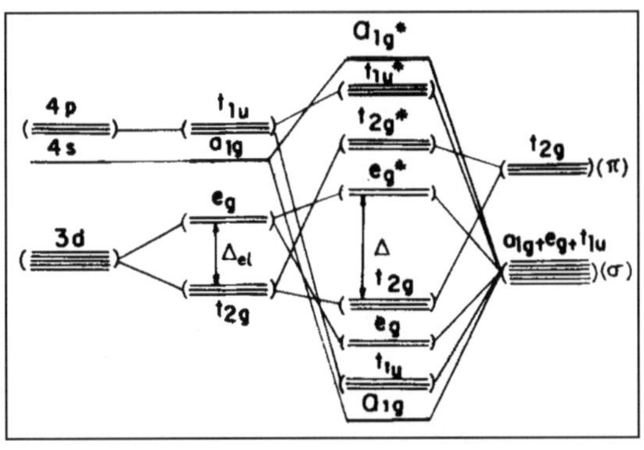

Electronic excited states for a coordination compound with Oh symmetry.

In the latter category three types of transitions are assigned: metal-to-ligand charge-transfer transitions (MLCT); ligand-to-metal charge-transfer and charge-transfer-to-solvent transitions (SCT).

The charge transfer bands in the electronic spectra of coordination compounds have been identified and assigned based on experimental data. Jorgensen attempted to work out theoretical models to explain the electron transfer spectra of Werner type coordination compounds. As far as the ligand field transitions are concerned, the assignment could be done based on the ligand field model more successfully with first row transition metal compounds whereas in higher transition metals the theoretical model does not adequately interpret the absorption spectra. Emission spectral studies in organic systems have provided valuable information on the relaxation processes of excited states. In fact many important excited state processes involving organic systems such as excelled or exciter formation have been extensively investigated using emission spectroscopy. Comprehensive models proposed by Rahm and Weller on excited state electron transfer reactions are based on fluorescence studies. In transition metal complexes, room temperature emission has been observed only in a small number of compounds for these studies to be of much assistance in the understanding of electronic excited states. In a few metal-complex systems where emission has "been observed considerable information has been derived on the nature of excited states and their deactivation pathways. In contrast to the organic molecules, coordination compounds which carry charge interact considerably strongly with the solvents or other environments. Interpretation of the excited processes becomes more complicated as the interaction between the solvent and the excited state is not understood well. In non-polar solvent most of the Werner type complexes are insoluble. In fact photochemical studies reveal information on the interaction of the excited states with solvent molecules. The nature of excited states in organometallic systems is not well understood as extensive pi-bonding occurs and most of the transitions seem to involve charge transfer processes. It is a long way before the excited state processes of organometallic complexes are understood based on the nature of the knowledge of the excited states.

Photochemical Reactions of Coordination Compounds

In the ground state, coordination compounds give rise to either ligand labilisation reactions or redox reactions. Ligand labilisation leads to aquation, anation or isomerisation reactions. Redox reactions lead to the oxidation or reduction of the metal centre with corresponding changes in the coordinating ligands. In general, photochemical reactions are different from the thermal reactions since the excited states have different electronic structure and as a consequence different geometry and other chemical properties from the ground state. Absorption of light by ligand field bands produces excited states restricted to metal centred orbitals. From the emission spectral studies depending upon the nature of the transitions involved (for instance $t_{2g} \rightarrow t_{2g}$ or $t_{2g} \rightarrow e_g$ in the octahedral complexes) extent of distortion of the geometry of the excited states from the ground state could be determined. This is schematically illustrated in figure for chromium(Ill) systems. Excitation of ligand field band produces thermal or anti-thermal products. Not much information is available on the rate of deactivation processes of ligand field excited states. For coordination compounds it appears that in the cases studied thus far the relaxation pathways and their characteristics differ markedly from those observed for organic systems. Charge-transfer excitation of complexes leads to either ligand labilisation or redox reactions.

Electron Transfer Reactions of Metal Complexes

Electron-transfer reactions mediated by metal ions have been recognised long ago. The importance of the electron transport in natural systems led Taube to investigate these reactions systematically.

Depending upon the ligands coordinated, the redox potentials of metal complexes vary dramatically as shown in table. This property is found utilised in many biological reactions where the same metal ion brings about different types of redox reactions depending upon the nature of the coordinated ligands. Thus, copper or iron is found to undergo large number of enzymatic reactions involving number of metalloenzymes. Using very simple systems electron-transfer reactions have been classified as inner sphere or outer-sphere reactions.

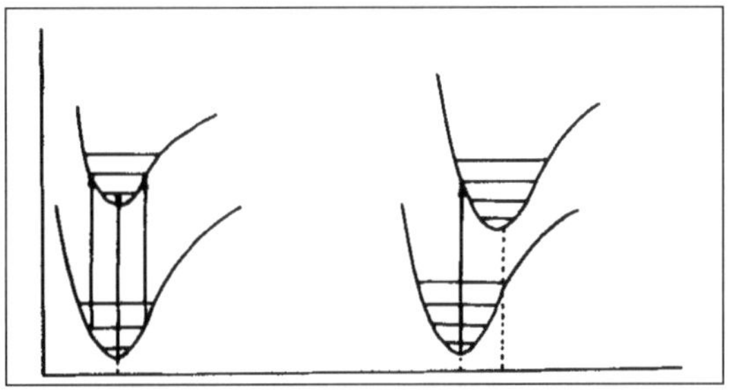

Geometry of the quartet and doublet excited states of chromium(III) complex.

Table: Redox potentials of metal complexes depending upon the coordinated ligand.

Couple	Redox potential V
$Fe(H_2O)_{6+}^{3+} + e \rightarrow Fe(H_2O)_6^{2+}$	0.77
$Cr(H2O)_6^{3+} + e \rightarrow Cr(H_2O)_6^{2+}$	-0.44
$Fe(bpy)_3^{3+} + e^- \rightarrow Fe(bpy)_3^{2+}$	1.05
$Ru(bpy)_3^{3+} + e^- \rightarrow ru(bpy)_3^{2+}$	1.26

Inner sphere,

$$Co(NH_3)_5Cl^{2+} + Cr(H_2O)_6^{2+} \rightarrow \left[Co(NH_3)_5ClCr(H_2O)_5\right]^{4+} \rightarrow$$
$$Co(NH_3)_5^{2+} + Cr(H_2O)_5Cl^{2+}$$

Outer sphere,

$$Co(NH_3)_5Cl^{2+} + Cr(bpy)_3^{2+} \rightarrow Co(NH_3)_5Cl^{+} + Cr(bpy)]_3^{3+}$$

For the same metal ions depending upon whether the reactions are inner sphere or outer sphere the reaction rate is changed. In the inner sphere reactions formation of a bridge between the oxidant and reductant makes energetics of the reaction different from the outer sphere reactions where the electron transfer is not mediated through a bridging ligand. Marcus has proposed a theoretical model for explaining the electron transfer reactions and derived a simplified relationship $k_{12} = (k_{11}k_{22}k_2 f)^{1/2}$ relating the rate constant for the cross-reaction k_{12} the equilibrium constant k_{12} and the self-exchange rate constant k_{11} for the oxidant and k_{22} for the reductant respectively; *f* is a factor close to unity. For some ground state reactions the electron transfer rate constants are given in table.

Bimolecular electron transfer reactions occur in three steps:

$$A + B = (AB),$$

$$(AB) = (A^+B^-),$$

$$(AB) \rightarrow A^+ + B^-.$$

The first equilibrium involves the formation of the precursor complex where no strong interactions occur between the reactants. In the second step the coordination spheres of the reactants are altered to facilitate electron transfer. In the inner sphere electron transfer reactions the bridge formation occurs in thi~ step. In the outersphere electron transfer reactions the coordination environment including the solvent sphere around the reactants are altered appropriate to the formation of the activated complex.

Reactions from the Charge-transfer-to-metal Excited States

Photochemistry of Cobah(lll) Complexes

Cobalt (III) ion with nd^6 configuration forms low spin octahedral complexes which are kinetically inert. Cobalt (III) ammines have well separated charge-transfer and ligand field bands, these complexes do not show any emission.

Table: Rate constants for electron transfer reactions in the ground state.

System	Rate constants (M^{-1} sec^{-1})
$Cr^{2+} + Cr(NH_3)_5Br^{2+}$	5.1×10^{-2}
$Fe^{2+} + Fe(dipy)_3^{3+}$	27×10^4
$Fe(CN)_6^{4-} + IrCl_6^{2-}$	1.9×10^6
$Cr^{2+} + Co(NH_3)_6^{3+}$	8.9×10^{-5}
$V^{2+} + Co(NH_3)_5Br^{2+}$	25
$Ru(NH_6)_6^{2+} + Co(NH_3)_5Br^{2+}$	1.6×10^3
$Eu^{2+} + Co(NH_3)_5OH_2^{3+}$	0.15

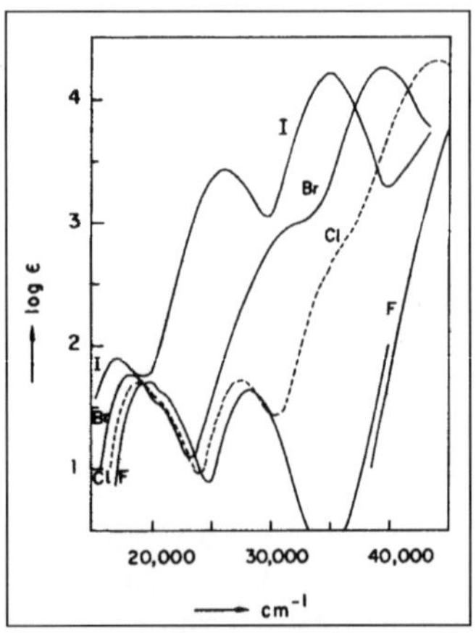

Charge-transfer transitions for $[Co(NH_3)_5X]^{2+}$, X = (Cl, Br, I) complexes.

The photochemistry of the ammines and acidoammines of cobalt(III) ion has been extensively investigated. Depending upon the ease of oxidation of the ligand, CTTM transitions shift as shown in figure for a series of acidoammine complexes. An empirical relationship is derived to relate the absorption band maximum and the electronegativity of the ligand. Irradiation in the CTTM band of the complex in aqueous solution in general results in the charge transfer to the metal centre which leads to the decomposition of the complex to give redox products. Quantum yields for cobalt(II) for the acidoammine complexes on irradiation at different wavelengths are given in table. The redox reaction occurs exclusively on irradiation in the charge-transfer band; irradiations in the ligand field bands do not produce cobalt(II). The proposed mechanism for the photoreduction of cobalt(III) ammines is given by the following scheme:

$$\left[Co^{III}(NH_3)_5 x\right] \xrightarrow{2+hv} {}^*\left(Co^{III}(NH_3)_5 x\right)^{2+} \text{(CTTM)}$$

$$\text{CTTM} \rightarrow \left(Co^{II}(NH_3)_5 \cdot x\right),$$

$$\left(Co^{II}(NH_3)_5 \cdot x\right) \rightarrow Co^{II}(NH_3)_5^{2+} + \cdot x,$$

$$Co^{II}(NH_3)_5^{2+} \xrightarrow{H+} Co(NH_3)_2^{2+} + 3NH_4^+ \; k > I0^6 \text{ sec}^{-1},$$

$$Co(NH_3)_2^{2+} \rightarrow Co(II) + 2NH_4^+.$$

Table: Quantum yield for the reduction of cobalt III amine complexes.

Complex	(nm) of excitation	Quantum yield
$Co(NH_3)]_6^{3+}$	254	0–16
$Co(en)_3^{3+}$	254	0–13
$Co(dien)_2^{3+}$	254	0–03
$Co(NH_3)_5C1^{2+}$	254	0-25
Cis $Co(en)_2Cl_2+$	254	0.06
$Co(NH_3)_5N_3^+$	254	0.20
$CO(EDTA)^-$	254	0.06

It has also been observed that with polydentate ligands the redox quantum yield decreases compared to the monodentate complexes. One of the important reactions occurring from the charge transfer excited states of cobalt(III) complex is the linkage isomerization reaction. For instance $(Co(NH_3)_5NO_2)^{2+}$ ion on irradiation in the CTTM band produces cobalt(II)and the linkage isomer.

$$\left(Co(NH_3)_5 NO_2\right)^{2+} \rightarrow Co(II) + \left(Co(NH_3)_5 ONO\right)^{2+}.$$

Considerable interest was shown to understand the process by which the nitrite isomer was formed. Scandola et al carried out the irradiation of the complex in waterglycerol mixtures and suggested that the increased formation of the linkage isomer with the concomitant decrease in cobalt(II) yield as the viscosity of the irradiated solution is due to the geminate radical recombination.

$$((NH_3)_5 CO^{11} \cdot NO_2),$$
$$\downarrow$$
$$(NH_3)_5Co^{III}\text{-}NO_2).$$

However investigations on the photochemical behavior of $Co(NH_3)_5 N_3^{2+}$ ion and *cis Co(en)*$_2$ $(NO_2)_2^+$ ion in highly viscous solutions containing synthetic high polymers or other alcohols did not show evidence for radical recombinations. It was suggested that the excited state relaxation process itself is affected by the nature of the medium which affects the quantum yields for products. More recently Natarajan and Radhakrishnan have studied the photoreaction of the complex *cis*-$Co(en)_2(NO_2)_2^+$ in neat acetonitrile and compared the behavior with that in aqueous solution. It is indicated that more the hydrophylic environment.of the excited state of cobalt(III) complex, cobaltous ion formation is facilitated. Several attempts to photosensitize cobalt(III) complexes have been made. Photosensitization of cobalt(III) complexes gives rise to cobalt(II) ions. However, the mechanism of photosensitization is still ambiguous. Although many early results have indicated that the photosensitization occurs by an energy transfer mechanism later it turned out that reexamination of the systems has shown that the quenching is due to a chemical mechanism in some cases and in other cases both physical and chemical mechanisms of sensitization take place. It is also of interest to investigate the extent of communication between different types of excited states. On irradiation in the ligand field bands the quantum yield for redox decomposition is negligibly small (10^{-3}) whereas excitation in the charge transfer region does not lead to aquation or photoanation reactions. It thus appears that higher energy charge transfer excited states do not relax passing through the low lying d-d excited states. The surfaces of and CTTM states do not seem to cross each other.

Photoredox Reactivity of CTTM States of the Complexes of Cu(II) and Other Metalions

Copper(II) forms kinetically labile complexes which on irradiation in aqueous solution produces copper(I) and ligand radical. The secondary reactions and the lability of copper(II) complexes make the systems more difficult to understand. Copper(II) acetate on irradiation in aqueous solution produces copper(I) and methyl radical. Flash photolysis studies indicate the formation of methyl radicals. Recently the photochemistry of copper(II) aminoacid complexes has been investigated using flash photolysis technique. The primary photochemical reaction involves the formation of copper(I) coordinated to the free radical. The copper I radical species are shown to initiate vinyl polymerisation reactions and steps involved in the photochemistry of copper(II)-aminoacid are shown below.

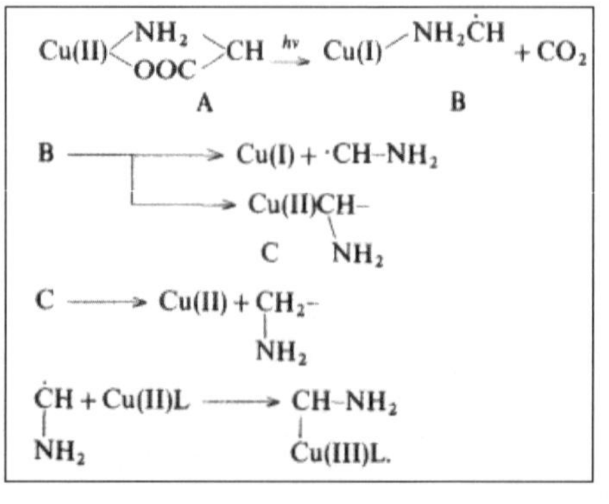

Studies of polymerisation kinetics have led to the calculation of primary quantum yields for the formation of the copper I coordinated free radical which is difficult to measure directly since secondary reactions enhance the extent of formation of copper(I). Copper(II) complexes with polypyridyl ligands undergo photoreactions to produce copper(I) with low quantum yields.

Photochemistry from CTTL States

In the case of metal complexes in which the metal is present in low oxidation states and the ligand electron-deficient or aromatic systems intense charge-transfer to ligand transitions have been observed in the visible or near uv region. Excitation ofc-'rTL bands of metal complexes produces CTTL excite states and the photochemistry of these systems has been of much interest in recent years. Ruthenium(II) ammines form complexes with pyridine and substituted pyridines. $Ru(NH_3)_5py^{2+}$ (py = pyridine) has absorption maximum in the visible region at 407 nm with molar absorptivity of 7500. The transition is assigned as MLCT and the peak shifts depending upon the substituents on pyridine. Corresponding ruthenium(III) complexes do not show this band supporting the MLCT assignment for the ruthenium(II) systems. On irradiation, the complex ion does not undergo redox reaction. On the other hand aquation of pyridine ligand takes place. Flash photolysis of the complex ion produces a transient and the transient has been proposed to have the structure.

The intermediate rearranges to assume a n-bonded pyridine with the metal which eventually is replaced by water. In the polypyridyl complexes of ruthenium(II), osmium(II) and iron(II) chargetransfer-to-ligand transitions have been observed in the visible region. In recent years the photochemistry of these complexes has been investigated in great detail. Tris(2,2'-dipyridine) ruthenium(II) complex, $Ru(bpy)_3^{2+}$ is a well-known compound which on excitation in the CTTL band produces an excited state with a lifetime of 0.5 sec and whose thermodynamic properties are quite different from those of the ground state. The excited state, $Ru(bpy)_3^{2+}$ serves as a good reducing agent.

Photochemistry from the Charge-transfer to Solvent Excited States

Excited states of molecules which involve the surrounding medium have been studied in detail. In particular interactions between the solute and sorvents are more in polar solvents and depending upon the polarity of the solvent changes in absorption spectra are observed. One of the well-known systems in which photochemistry from the CTTM excited state occurs is potassium hexacyanoferrite(II). The photochemical reaction is given by,

$$Fe(CN)_6^{4-} \xrightarrow[H_2O]{hv} Fe(CN)_6^{3-} + e^-(H_2O)$$

and hydrated electron has been detected by scavenging experiments and in absorption spectra.

Excited States as Electron Transfer Reagents

While the redox reactions have been studied extensively the excited state electron transfer reactions have been discovered only in recent years. Pioneering work on excited state electron transfer reactions has been carried out by Weller and his coworkers in the case of organic molecules. Electron transfer reactions of excited states of transition metal complexes are of particular interest since they are promising for solar energy conversion to fuels or electricity and also for the production of metal ions in unusual oxidation states. In general, excited state electron transfer reactions with a quencher may be given by the scheme.

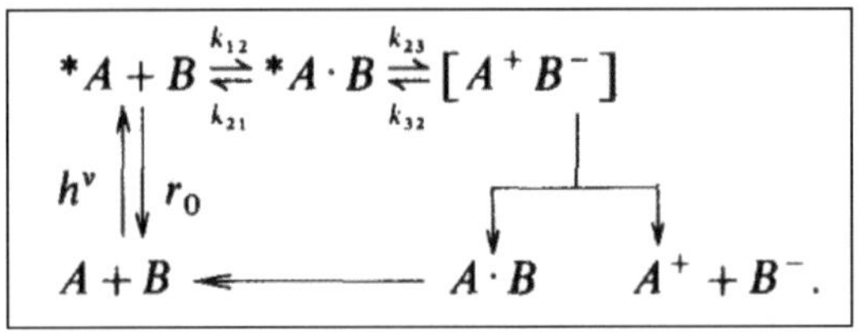

The rate constant for the excited state electron transfer reaction is given by,

$$kq^{obs} = \frac{k_{12}}{1 + \frac{k_{12}}{\Delta v \cdot k_{30}} \exp(G^{*}_{23} / RT) + \exp(\Delta G_{23} / RT)}.$$

One of the most important bimolecular redox reactions involving an excited state of a coordination compound is,

$$*Ru(bpy)_3^{2+} + Fe(III) \rightarrow Fe(II) + Ru(bpy)_3^{3+}.$$

The rate constant for the reaction is determined to be 10^9 M^{-1} sec^{-1} in aqueous acidic solution at 25° the reaction products are of high energy and they recombine to produce the ground states of the starting materials. The energetics of the $Ru(bpy)_3^{2+}$ -Fe(II) system in different oxidation states is given in figure.

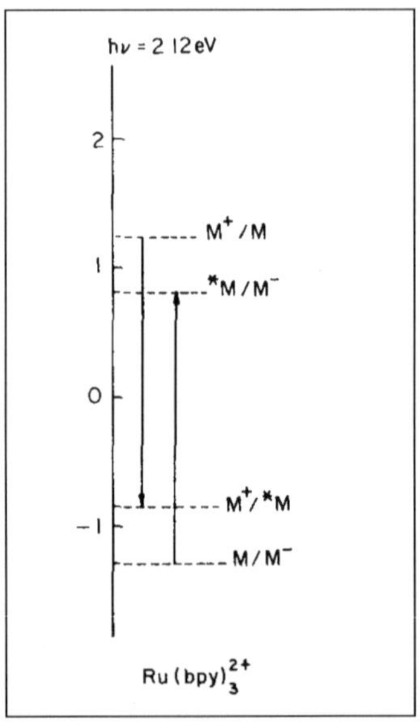

Redox potentials (Volts)of the ground and excited states of $Ru(bpy)_3^{2+}$ ion.

When methylviologen is used as a quencher of $Ru(bpy)_3^{2+}$ the following reactions take place.

$$Ru(bpy)_3^{2+} + PQ^{2+} \xrightarrow{kq} {}^*Ru(bpy)_3^{3+} + PQ^+,$$

$$PQ^+ + Ru(bpy)_3^{3+} \rightarrow PQ^{2+} + Ru(bpy)_3^{2+}.$$

A plot of the electron transfer quenching constant versus reduction potential of the quenchers clearly shows the point at which electron transfer quenching becomes endothermic. From the ground state potential in the same medium, $E_{1/2}$, for $Ru(bpy)_3^{3+}/Ru(bpy)_3^{2+} = 1.29$ V) it can be seen that photoexcitation increases the oxidising power by 2. 10 V, a value very close to the spectroscopic excited state energy (2.18 V) of $^*Ru(bpy)_3^{2+}$.

Europium(II) aquo ions can reduce the excited state of $Ru(bpy)_3^{2+}$ to $Ru(bpy)_3^{3+}$ to produce europium(III) ions.

References

- Coordination-compound, science: britannica.com, Retrieved 19 April, 2019
- Chelate, science: britannica.com, Retrieved 9 February, 2019
- IUPAC, Compendium of Chemical Terminology, 2nd ed. (the "Gold Book") (1997). Online corrected version: (2006–) "bridging ligand". doi:10.1351/goldbook.B00741
- Fundamentals-of-Chirality, Chirality, Organic-Chemistry: libretexts.org, Retrieved 10 August, 2019
- Kubas, Gregory J. (2001). Metal Dihydrogen and σ-Bond Complexes - Structure, Theory, and Reactivity (1 ed.). New York: Kluwer Academic/Plenum Publishers. ISBN 978-0-306-46465-2. LCCN 00059283
- Ligand-substitution, Fundamentals-of-Organometallic-Chemistry, Organometallic-Chemistry, Inorganic-Chemistry: libretexts.org, Retrieved 6 January, 2019
- Mechanism-of-Ligand, Ligand-Substitution-in-Coordination-Complexes, Reactivity, Reactions, Organic-Chemistry: libretexts.org, Retrieved 16 March, 2019
- Anusavice KJ (2012-09-27). "Chapter 12: Bonding and Bonding Agents". Phillips' Science of Dental Materials (12th ed.). Elsevier Health. pp. 257–268. ISBN 978-1-4377-2418-9. OCLC 785080357

5

Inorganic Reactions

Reactions in inorganic chemistry are divided into a few broad categories, namely, combination reactions, decomposition reactions, replacement reactions and combustion reactions. The chapter closely examines these key reactions in inorganic chemistry to provide an extensive understanding of the subject.

Combination Reactions

A combination reaction is a reaction in which two or more substances combine to form a single new substance. Combination reactions can also be called synthesis reactions. The general form of a combination reaction is:

$$A+B \rightarrow AB$$

One combination reaction is two elements combining to form a compound. Solid sodium metal reacts with chlorine gas to produce solid sodium chloride:

$$2Na(s)+Cl_2(g) \rightarrow 2NaCl(s)$$

In order to write and balance the equation correctly, it is important to remember the seven elements that exist in nature as diatomic molecules ($H_2, N_2, O_2, Cl_2, Br_2$, and I_2).

One sort of combination reaction that occurs frequently is the reaction of an element with oxygen to form an oxide. Metals and nonmetals both react readily with oxygen under most conditions. Magnesium reacts rapidly and dramatically when ignited, combining with oxygen from the air to produce a fine powder of magnesium oxide:

$$2Mg(s)+O_2(g) \rightarrow 2MgO(s)$$

Sulfur reacts with oxygen to form sulfur dioxide:

$$S(s)+O_2(g) \rightarrow SO_2(g)$$

When nonmetals react with one another, the product is a molecular compound. Often, the non-metal reactants can combine in different ratios and produce different products. Sulfur can also combine with oxygen to form sulfur trioxide:

$$2S(s)+3O_2(g) \rightarrow 2SO_3(g)$$

Transition metals are capable of adopting multiple positive charges within their ionic compounds. Therefore, most transition metals are capable of forming different products in a combination reaction. Iron reacts with oxygen to form both iron (II) oxide and iron (III) oxide.

$$2Fe(s) + O_2(g) \rightarrow 2FeO(s)$$

$$4Fe(s)+3O_2(g) \rightarrow 2Fe_2O_3(s)$$

Combination reactions can also take place when an element reacts with a compound to form a new compound composed of a larger number of atoms. Carbon monoxide reacts with oxygen to form carbon dioxide according to the equation:

$$2CO(g)+O_2(g) \rightarrow 2CO_2(g)$$

Two compounds may also react to form a more complex compound. A very common example is the reactions of oxides with water. Calcium oxide reacts readily with water to produce an aqueous solution of calcium hydroxide:

$$CaO(s)+H_2O(l) \rightarrow Ca(OH)_2(aq)$$

Sulfur trioxide gas reacts with water to form sulfuric acid. This is an unfortunately common reaction that occurs in the atmosphere in some places where oxides of sulfur are present as pollutants. The acid formed in the reaction falls to the ground as acid rain:

$$SO_3(g)+H_2O(l) \rightarrow H_2SO_4(aq)$$

Acid rain has severe consequences on both natural and manmade objects. Acid rain degrades marble statues like the one on the left (A). The trees in the forest on the right (B) have been killed by acid rain.

Combination of Elements to make Ionic Substances

An elemental metal and an elemental non-metal react to make an ionic substance that is neutral and has each ion in its correct charge state or valence. For instance,

$$2Na(s) + Cl_2(g) \rightarrow 2NaCl(s)$$

$$2Mg(s) + O_2(g) \rightarrow 2MgO(s)$$

$$2Al(s) + 3O_2(g) \rightarrow Al_2O_3(s)$$

If the metal is a transition metal, it will be much harder to predict the correct charge on the metal ion in the ionic compound.

Under what circumstances do these reactions happen? Often, an elemental metal and non-metal "want" to make an ionic compound, because this is a more stable state (think about a heavy ball on a table: It can easily roll to the ground, where it has less potential energy, so the table isn't a stable state; if the heavy ball is in a small hole in the ground, it can't easily move, and if it did, it would have more potential energy, so the hole is a stable, low energy state). However, that doesn't necessarily mean the reaction will just happen on its own. That depends on how easily the reaction can happen (think about a place you want to go, but don't go because traveling there is very inconvenient).

A ball being held above a ledge has more potential energy so it is in a less stable state than if it were at the bottom of the ledge.

For instance, the alkali metals and the halogens react pretty easily, so they will often react even without a "push." Oxygen is very reactive, which is why things burn, but you have to light them on fire to get them started. This is good, because otherwise we would burn in air at room temperature. Many of these elemental combination reactions might need a high temperature to get started, even if they want to happen. It won't be hard to remember that alkalis, alkaline earth metals and halogens react easily, because this is why they are very hard to find in elemental form! Oxygen and nitrogen are very abundant in elemental form because it is hard for them to react even if they want to. Nitrogen in particular reacts only with lithium metal and a few complicated compounds at room temperature, although it will react with many other elements at high temperatures. Most metals aren't found in elemental form in nature (except for ones that don't want to react, like gold), but if you find them in elemental form in your house, then probably they don't react easily.

Combination of Elemental Non-metals into Covalent Compounds

These reactions involve elemental forms of elements like H, C, N, O, Cl, S, P, etc. It will often be hard to predict the product because these elements can often combine in different ratios (this is where the law of multiple proportions comes from). You can always expect that H will have a valence of 1, and O will usually have a valence of 2. Many of these reactions will happen quickly if you get them started with a little heat, especially if oxygen or a halogen is involved. Otherwise, they might happen very slowly or not at all except under special circumstances. Some examples:

$$C(s)+O_2(g)\rightarrow CO_2(g)(fast, once\,lit)$$
$$N_2(g)+3H_2(g)\rightarrow 2NH_3(g)(usually\,very\,slow).$$

Basic Anhydrides

Basic anhydrides are compounds that turn into a base (a hydroxide salt) when you add water. They are metal oxides. Here's an example:

$$CaO(s)+H_2O(l)\rightarrow Ca(OH)_2(aq).$$

If the metal is an alkali or alkaline earth, the reaction probably happens quickly and produces a lot of heat. If the metal is a transition metal, the reaction might not happen so easily or at all.

Acid Anhydrides

Acid anhydrides are compounds that turn into an acid when you add water. They are non-metal oxides. These are a little more complicated than basic anhydrides, so don't worry too much about them right now. Here's an example:

$$SO_3(g)+H_2O(l)\rightarrow H_2SO_4(aq)$$

Other Combination Reactions

There are many other circumstances in which a combination reaction could happen. The types listed here are the simple ones that are good to know in the beginning.

Examples for combination reaction:

1. $Na_2 + H_2O$ --> 2NaOH

   Here sodium oxide is combined with water gives sodium hydroxide.

2. 4Li + O_2 --> $2Li_2O$

   Here lithium is combined with oxygen to form lithium oxide.

3. 2Mg + O_2 --> 2MgO

   Here magnesium is combined with oxygen to form magnesium oxide.

4. $H_2 + Cl_2$ --> 2HCL

   Here hydrogen is combined with chlorine to form hydrochloride.

5. NH_3 + HCl --> NH_4Cl

   Here colorless ammonia is added to colorless hydrogen chloride gives ammonia chloride.

6. $2H_2 + O_2$ --> $2H_2O$

   Here hydrogen is burned in the air or oxygen produces water.

7. S + O_2 --> SO_2

   Here sulphur is burned in the air to produce sulphur dioxide.

8. Ba + F_2 --> Ba F_2

   Here barium metal and fluorine gas is combined to produce salt barium fluoride which is highly exothermic reaction.

9. MgO + CO_2 --> $MgCO_3$

   Here magnesium oxide is combined with carbon to produce magnesium carbon dioxide.

10. 2Al + 3Br --> $2AlBr_3$

    Here aluminum is combined with bromine which produces 3 + aluminum iron and 1+bromine iron.

Other Reactions

Chemical reactions are of the following six types:

1. Combustion reaction: It is a type of reaction in which oxygen combines with the other compound and forms carbon dioxide (CO_2) and water (H_2O).

   $C_{10}H_8 + 12\ O_2$ -----------> $10\ CO_2 + 4\ H_2O$

   $CH_4 + 2\ O_2$ ---------------> $CO_2 + 2\ H_2O$

2. Synthesis Reaction: It is a type of chemical reaction in which two or more simple substances combine to form one single complex substance.

   $C + O_2$ ----------> CO_2

   $CaO + H_2O$ -----------> $Ca(OH)_2$

3. Decomposition Reaction: It is a type of chemical reaction in which a complex substance breaks down to simpler substances. It is opposite of synthesis reaction.

   $2\ N_2O_5$ ----------> $4\ NO_2 + H_2O$

   $2\ KClO_3$ -------------> $2\ KCl + 3\ O_2$

4. Single displacement reaction: It is a chemical reaction in which an element displaces another element in a compound.

   $Mg + 2\ HCl$ -----------> $MgCl_2 + H_2$

   $Fe_2O_3 + Zn$ ------------> $Zn_2O_3 + 2\ Fe$

5. Double displacement reaction: This is a type of chemical reaction in which the anions and cations in the reactants interchange their positions and form two different compounds.

   $2\ NaCl + H_2SO_4$ -------------> $Na_2SO_4 + 2HCl$

   $AgNO_3 + KCl$ ---------------> $AgCl + KNO_3$

6. Acid - Base reaction: It is a type of double displacement reaction in which acid and base react to form salt and water.

   $HCl + NaOH$ -------------> $NaCl + H_2O$

   $H_2SO_4 + 2\ NaOH$ ----------> $Na_2SO_4 + 2\ H_2O$

7. Isomerisation: It a type of chemical reaction in which a chemical compound undergoes a structural rearrangement without any change in its net atomic composition.

8. Disproportionation: It's a type of chemical reaction in which a single reactant forms two products with different oxidation states.

   Pb+2 ------------> Pb + Pb +4.

   The products lead has 0 and +4 oxidation states.

Decomposition Reactions

A decomposition reaction is a type of chemical reaction in which one reactant yields two or more products.

The general form for a decomposition reaction is,

$$AB \rightarrow A + B$$

Decomposition reactions are also known as analysis reactions or chemical breakdowns. The opposite of this type of reaction is a synthesis, in which simpler reactants combine to form a more complex product.

You can recognize this type of reaction by looking for a single reactant with multiple products.

In certain circumstances, decomposition reactions are undesirable. However, they are intentionally caused and analyzed in mass spectrometry, gravimetric analysis, and thermo gravimetric analysis.

Decomposition Reaction Examples

The electrolysis of water into oxygen and hydrogen gas is an example of a decomposition reaction:

$$2\,H_2O \rightarrow 2\,H_2 + O_2$$

Another example is the decomposition of potassium chloride into potassium and chlorine gas:

$$2\,KCl_{(s)} \rightarrow 2\,K_{(s)} + Cl_{2(g)}$$

Most decomposition reactions require an input of energy in the form of heat, light, or electricity.

Binary compounds are compounds composed of just two elements. The simplest kind of decomposition reaction is when a binary compound decomposes into its elements. Mercury (II) oxide, a red solid, decomposes when heated to produce mercury and oxygen gas.

$$2HgO(s) \rightarrow 2Hg(l) + O_2(g)$$

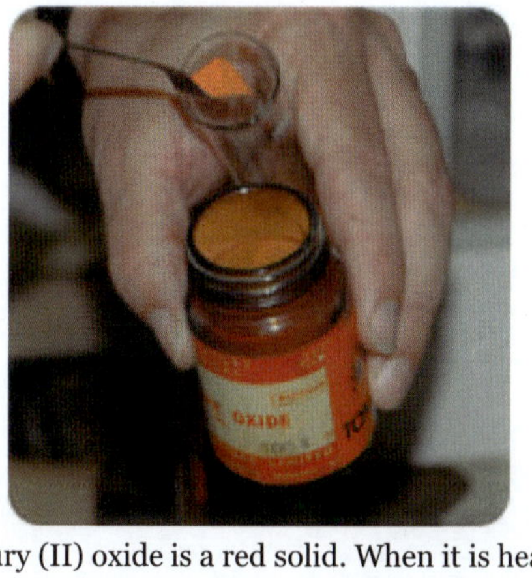

Mercury (II) oxide is a red solid. When it is heated, it decomposes into mercury metal and oxygen gas.

A reaction is also considered to be a decomposition reaction even when one or more of the produces are still compounds. A metal carbonate decomposes into a metal oxide and carbon dioxide gas. For example, calcium carbonate decomposes into calcium oxide and carbon dioxide.

$$CaCO_3(s) \rightarrow CaO(s) + CO_2(g)$$

Metal hydroxides decompose on heating to yield metal oxides and water. Sodium hydroxide decomposes to produce sodium oxide and water.

$$2NaOH(s) \rightarrow Na_2O(s) + H_2O(g)$$

Some unstable acids decompose to produce nonmetal oxides and water. Carbonic acid decomposes easily at room temperature into carbon dioxide and water.

$$H_2CO_3(aq) \rightarrow CO_2(g) + H_2O(l)$$

Replacement Reaction

A replacement reaction occurs when elements switch places in compounds. This type of reaction involves ions (electrically charged versions of atoms) and ionic compounds. These are compounds in which positive ions of metal and negative ions of a nonmetal are held together by ionic bonds. Generally, a more reactive element replaces an element that is less reactive, and the less reactive element is set free from the compound. There are two types of replacement reactions: single and double.

Single Replacement Reactions

A single replacement reaction, sometimes called a single displacement reaction, is a reaction in which one element is substituted for another element in a compound. The starting materials are always pure elements, such as a pure zinc metal or hydrogen gas, plus an aqueous compound. When a replacement reaction occurs, a new aqueous compound and a different pure element will be generated as products.

The general pattern of a single replacement reaction is shown below,

$$AB(aq) + C \rightarrow A + CB(aq)$$

$$\downarrow \quad \downarrow$$

Pure elements

We can see that A is replaced by C in compound AB to make a new compound CB and elemental A. Another thing you might notice is that A start out as an ion in solution but appears in its elemental form on the product side. The reactant C does the opposite: it starts out in its elemental form on the reactant side, but it ends up as an ion in aqueous solution as part of the compound CB (aq).

Let's try to make sense of that definition with an example reaction.

$$\underset{\substack{clear, colorless\\ solution}}{\underset{\downarrow}{AgNO_3(aq)}}+Cu(s) \rightarrow \ ?$$

Color change plus the formation of precipitate?
Looks like a chemical reaction.

We start with a clear, colorless solution of silver(I) nitrate, then we drop in some shiny copper wire. The solution turns aquamarine blue, and the copper wire starts to look grey and fuzzy.

Determining the Products of Single Replacement Reactions

If we are trying to figure out whether a single displacement reaction will occur:

1. In proposed reaction two elements that might swap their places:

In general, elements that form anions can replace the anion in a compound, and elements that form cations can replace the cation in a compound. The following guidelines can be used to determine what kind of ions a given element might form.

- Metals will usually form cations - This includes groups 1 and 2, some of group 13 and 14 elements, and the transition metals.
- The common non-metals in single replacement reactions are the group 17 elements, which generally form anions with a 1- charge.
- Hydrogen usually forms the cation H^+ in a single replacement reaction.

In our reaction with copper metal and aqueous silver (I) nitrate, the copper metal will likely react to form copper cations because it is a transition metal. The copper cations can replace the silver cations in the compound $AgNO_3(aq)$ to form a new compound.

2. Formation as a product by new compound:

Once we know what element might be replaced in our ionic compound, we can predict the products that might be formed. In this example, the silver atoms in $AgNO_3(aq)$ can be replaced by copper to form $Cu(NO_3)_2(aq)$.

In the process, elemental silver, Ag(*s*) would also form as a product. We can write out the full—and balanced—reaction as follows:

$$2AgNO_3(aq) + Cu(s) \rightarrow Cu(NO_3)_2(aq) + 2Ag(s)$$

It turns out that aqueous solution of $Cu(NO_3)_2$ re blue-green, which explains the solution's color change. The grey fuzz growing on the copper would be from silver metal precipitating out on the surface of the wire.

Predicting if a Single Replacement Reaction will Occur

Once we know which elements might get swapped in our single displacement reaction, we can predict whether the reaction will occur based on knowledge of the relative reactivates of the two elements—elements C and A in the generic pattern above, or copper and silver in our example reaction. If element C is more reactive than element A, then C will replace A in a compound. If element C is less reactive than element A, then there will be no reaction.

The reactivity series—also called the activity series—ranks elements in order of their reactivity for certain types of reactions, including single replacement reactions. The more reactive elements will replace the less reactive elements in the reactivity series, but not the other way around. There are separate rankings for elements that form cations and elements that form anions.

For elements that tend to gain electrons to form anions, the order of reactivity from most reactive to least reactive goes as follows:

Most reactive $F_2 > Cl_2 > Br_2 > I_2$ Least reactive

For these elements, you can also look at their arrangement on the periodic table—group 17—to remember the order of reactivity. The higher the element's position in the column, the more reactive it will be. Based on this activity series, we would predict thatBr_2 would replace I_2 in a single replacement reaction, but Br_2 would not react with a compound containing fluoride ions.

For the cation forming elements, the reactivity series is longer, and the trends are not as straightforward. You can see an example of the reactivity series for cations below.

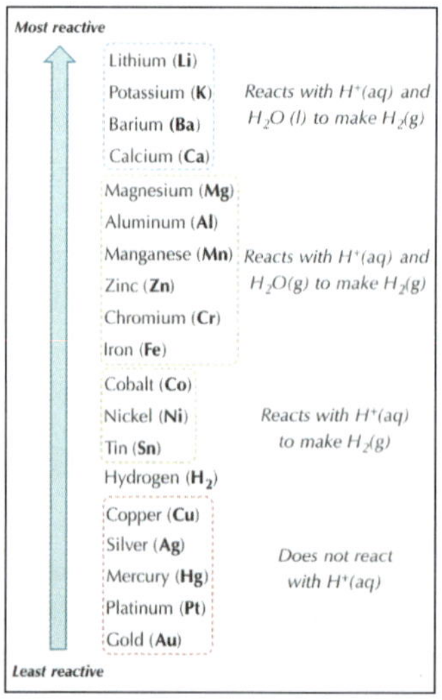

Reactivity is pretty complicated! After all, there are lots of different kinds of reactions, so what kind of reactivity are we really ranking here? Some properties that are taken into account in the reactivity series include reactivity with water and acids, as well as how readily an element loses electrons to form cations. As a result of the different ways reactivity can be defined, however, you might see some elements ranked in a different order.

The process of using the reactivity series is the same for both cations and anions: More reactive elements will replace less reactive elements in a compound.

Let's think back to our experiment combining $AgNO_3(aq)$ and copper wire. In the cation reactivity series, we see that copper is ranked higher than silver, so we would expect copper to be more reactive than silver in a single replacement reaction. Therefore, we would predict that Ag^+ Would get replaced by Cu^{2+} in a compound, which matches our results.

Predicting the Products of a Single Replacement Reaction

Let's consider the following reaction,

$$AlPO_4(aq) + Mg(s) \rightarrow$$

The first question we might ask is what element Mg might be replacing in the compound $AlPO_4$. Al is a metal that usually forms cations with a charge of 3+. We can verify this because $AlPO_4$ is neutral and phosphate has a 3- charge, so the aluminum cation must have a 3+ charge. Since Mg is also a metal that forms cations, we might expect Mg to replace the metal Al in our compound. If we check the cation reactivity series, we see that magnesium is more reactive than aluminum, so we predict the single replacement reaction will occur.

What products do we expect from this single replacement reaction? We expect to form elemental Al(s) and the new ionic compound $Mg_3(PO_4)_2$.

That gives the following reaction,

$$AlPO_4(aq) + Mg(s) \rightarrow Al(s) + Mg_3(PO_4)_2(aq)$$

We are not quite done though since our reaction is not currently balanced. We can fix this by multiplying $AlPO_4$ on the reactant side by two Mg(s) by three, and Al(s) on the product side by two. This gives us our final balanced equations,

$$2AlPO_4(aq) + 3Mg(s) \rightarrow 2Al(s) + Mg_3(PO_4)_2(aq)$$

The cup shown provides an example of tarnish, a chemical reaction caused when silver metal reacts with hydrogen sulfide gas produced by some industrial processes or as a result of decaying animal or plant materials,

$$2Ag + H_2S \rightarrow Ag_2S + H_2$$

The tarnish can be removed using a number of polishes, but the process also removes a small amount of silver along with the tarnish.

A tarnished silver cup.

$$A + BC \rightarrow AC + B$$

In this general reaction, element A is a metal and replaces element B, also a metal, in the compound. When the element that is doing the replacing is a nonmetal, it must replace another nonmetal in a compound, and the general equation becomes:

$$Y + XZ \rightarrow XY + Z$$

Where Y is a nonmetal and replaces the nonmetal Z in the compound with X.

Metal Replacement

Magnesium is a more reactive metal than copper. When a strip of magnesium metal is placed in an aqueous solution of copper (II) nitrate, it replaces the copper. The products of the reaction are aqueous magnesium nitrate and solid copper metal.

$$Mg(s) + Cu(NO_3)_2(aq) \rightarrow Mg(NO_3)_2(aq) + Cu(s)$$

This subcategory of single-replacement reactions is called a metal replacement reaction because it is a metal that is being replaced (copper).

Hydrogen Replacement

Many metals react easily with acids and when they do so, one of the products of the reaction is hydrogen gas. Zinc reacts with hydrochloric acid to produce aqueous zinc chloride and hydrogen.

$$Zn(s) + 2\,HCl(aq) \rightarrow ZnCl_2(aq) + H_2(g)$$

In a hydrogen replacement reaction, the hydrogen in the acid is replaced by an active metal.

Zinc metal reacts with hydrochloric acid to give off hydrogen gas in a single-displacement reaction.

Some metals are so reactive that they are capable of replacing the hydrogen in water. The products of such a reaction are the metal hydroxide and hydrogen gas. All group 1 metals undergo this type of reaction. Sodium reacts vigorously with water to produce aqueous sodium hydroxide and hydrogen.

$$2\,Na(s) + 2H_2O(l) \rightarrow 2\,NaOH(aq) + H_2(g)$$

Sodium metal reacts vigorously with water, giving off hydrogen gas. A large piece of sodium will often generate so much heat that the hydrogen will ignite.

Halogen Replacement

The element chlorine reacts with an aqueous solution of sodium bromide to produce aqueous sodium chloride and elemental bromine.

$$Cl_2(g) + 2\,NaBr(aq) \rightarrow 2NaCl(aq) + Br_2(l)$$

The reactivity of the halogen group (group 17) decreases from top to bottom within the group. Fluorine is the most reactive halogen, while iodine is the least. Since chlorine is above bromine, it is more reactive than bromine and can replace it in a halogen replacement reaction.

Double Displacement Reactions

Double replacement reactions—also called double displacement, exchange, or metathesis reactions—occur when parts of two ionic compounds are exchanged, making two new compounds. The overall pattern of a double replacement reaction looks like this,

$$A^+B^- + C^+D^- \rightarrow A^+D^- + C^+B^-$$

You can think of the reaction as swapping the cations or the anions, but not swapping both since you would end up with the same substances you started with. The solvent for a double replacement reaction is usually water, and the reactants and products are usually ionic compounds—but they can also be acids or bases.

Here is an example of a double replacement reaction,

$$BaCl_2(aq) + Na_2SO_4(aq) \rightarrow BaSO_4(s) + 2NaCl(aq)$$

In this example, the cations are Ba^{2+} and Na^{+} and the anions are cl⁻ and SO_4^{2-}. If we swap the anions, or cations, we get as our products $BaSO_4$ and NaCl.

Precipitation and Neutralization Reactions

Identifying double replacement reactions is usually fairly straightforward once you can recognize the pattern. Predicting whether the reaction will occur can be trickier; it helps to be able to recognize some common types of double replacement reactions.

A precipitation reaction is when two aqueous ionic compounds form a new ionic compound that is not soluble in water. One example is the reaction between lead (II) nitrate and potassium iodide. Both compounds are white solids that can be dissolved in water to make clear, colorless solutions. When you combine the two clear solutions, you get the following reaction:

$$\mathrm{Pb(NO_3)_2}(aq) + 2\,\mathrm{KI}(aq) \rightarrow 2\,\mathrm{KNO_3}(aq) + \mathrm{PbI_2}(s)$$

We made a beautiful golden solid from two clear solutions. In real life, your reaction flask might look something like the picture below.

Combining aqueous solutions of lead (II) nitrate and potassium iodide results in the formation of insoluble lead (II) iodide, a yellow solid.

The insoluble product compound is called the precipitate. The solvent and soluble components of the reaction are called the supernatant or supinate. We can use solubility rules to predict whether a precipitation reaction will take place. The formation of a solid precipitate is the driving force that makes the reaction proceeds in the forward direction.

Supernatant

Our supernatant is everything that is not the precipitate, so it is made up of our solvent H_2O and KNO_3 (*aq*), which can also be written as the dissociated ions K^{+}(*aq*) and NO_3^{-} (*aq*).

Neutralization reactions are a type of double replacement reaction that occurs between an acid and a base. The following is an example of a neutralization reaction,

$$\mathrm{HF}(aq) + \mathrm{NaOH}(aq) \rightarrow \mathrm{H_2O} + \mathrm{NaF}(aq)$$
$$acid + base \rightarrow \mathrm{H_2O} + salt$$

An aqueous neutralization reaction generally produces water and a new ionic compound, also called a salt. The hardest part of identifying a neutralization reaction is recognizing that you have an acid and a base for the reactants. Once you know you have a neutralization reaction, you can generally predict the reaction will occur in the forward direction as long as you have a strong acid and/or a strong base as a reactant.

A fun neutralization reaction that you may have tried is the combination of baking soda—sodium bicarbonate $NaHCO_3$ and vinegar—mostly water with acetic acid, $CH_3COOH(aq)$ — which produces carbonic acid - H_2CO_3 – and sodium acetate –$NaCH_3COO$, If you have tried this reaction at home, you probably remember a lot of fizzing because the neutralization reaction is accompanied by a gas-producing reaction, where the carbonic acid decomposes into carbon dioxide gas—bubbles—and water.

The double replacement reactions can be written as molecular, complete ionic or net ionic equations.

Predicting and Balancing a Double Replacement Reaction

Let's take a look at an example where we don't know the products:

$$H_2SO_4(aq) + Ba(OH)_2(aq) \rightarrow$$

First, we can identify the cations and anions that will get swapped. The cations are H^+ and Ba^{2+} and the anions are SO_4^{2-} and OH^-. Swapping anions gives the products H_2O and $BaSO_4$

$$H_2SO_4(aq) + Ba(OH)_2(aq) \rightarrow H_2O + BaSO_4$$

We can see that our double replacement reaction is also a neutralization reaction since we are reacting sulfuric acid, a strong acid, with barium hydroxide, a strong base. What is the state of the product barium sulfate? If we check our solubility rules, we see that barium sulfate is insoluble and should precipitate out of solution. That means our reaction is a precipitation reaction, too! We can also include that information in our equation by adding the symbol (*s*) after the $BaSO_4$.

We aren't quite done yet, though. Our reaction is not balanced since we have unequal numbers of hydrogens and oxygens on both sides of the arrow. We can fix this by multiplying H_2O by 2 to give our final, balanced molecular equation:

$$H_2SO_4(aq) + Ba(OH)_2(aq) \rightarrow 2H_2O + BaSO_4(s)$$

Double replacement reactions have two ionic compounds that are exchanging anions or cations. Precipitation reactions and neutralization reactions are two common types of double replacement reactions. Precipitation reactions produce an insoluble product from two aqueous reactants, and you can identify a precipitation reaction using solubility rules. Neutralization reactions occur when the reactants are an acid and a base, and neutralization reactions are usually favorable as long as the reaction involves a strong acid and a strong base.

Formation of a Precipitate

A precipitate forms in a double-replacement reaction when the cations from one of the reactants combine with the anions from the other reactant to form an insoluble ionic compound. When aqueous solutions of potassium iodide and lead (II) nitrate are mixed, the following reaction occurs.

$$2\,KI(aq) + Pb(NO_3)_2(aq) \rightarrow 2\,KNO_3(aq) + PbI_2(s)$$

There are very strong attractive forces that occur between Pb^{2+} and I^- ions and the result is a brilliant yellow precipitate. The other product of the reaction, potassium nitrate, remains soluble.

Formation of lead iodide precipitate.

Formation of Gas

Some double-replacement reactions produce a gaseous product which then bubbles out of the solution and escapes into the air. When solutions of sodium sulfide and hydrochloric acid are mixed, the products of the reaction are aqueous sodium chloride and hydrogen sulfide gas.

$$Na_2S(aq) + 2\,HCl(aq) \rightarrow 2\,NaCl(aq) + H_2S(g)$$

Formation of Molecular Compound

Another kind of double-replacement reaction is one that produces a molecular compound as one of its products. Many examples in this category are reactions that produce water. When aqueous hydrochloric acid is reacted with aqueous sodium hydroxide, the products are aqueous sodium chloride and water.

$$HCl(aq) + NaOH(aq) \rightarrow NaCl(aq) + H_2O(l)$$

Occasionally, a reaction will produce both a gas and a molecular compound. The reaction of a sodium carbonate solution with hydrochloric acid produces aqueous sodium chloride, carbon dioxide gas, and water.

$$Na_2CO_3(aq) + 2\,HCl(aq) \rightarrow 2\,NaCl(aq) + CO_2(g) + H_2O(l)$$

Acid-base Reactions

An acid-base reaction is a chemical reaction that occurs between an acid and a base. Several concepts exist which provide alternative definitions for the reaction mechanisms involved and their application in solving related problems. Despite several similarities in definitions, their importance becomes apparent as different methods of analysis when applied to acid-base reactions for gaseous or liquid species, or when acid or base character may be somewhat less apparent.

Common Acid-base Theories

Lavoisier

Historically, the first of the scientific concepts of acids and bases was provided by the French-chemist Antoine Lavoisier, circa 1776. Lavoisier's knowledge of strong acids was mainly restricted to oxyacids, which tend to contain central atoms in high oxidation states surrounded by oxygen, such as HNO_3 and H_2SO_4, and he was not aware of the true composition of the hydrohalic acids, HCl, HBr, and HI. From his limited knowledge, he defined acids in terms of their content of *oxygen,* and he named oxygen from Greek words meaning "acid-former".

The Lavoisier definition was held as absolute truth for over 30 years, in which Davy proved the lack of oxygen in hydrogen sulfide (H_2S), hydrogen telluride (H_2Te), and the hydrohalic acids.

Liebig

Around 1838, Justus von Liebig proposed a new definition of acids, based on his extensive works on the chemical composition of organic acids. This finished the doctrinal shift from oxygen-based acids to hydrogen-based acids, started by Davy. According to Liebig, an acid is a hydrogen-containing substance in which the hydrogen could be replaced by a metal. Liebig's definition, while completely empirical, remained in use for almost 50 years until the adoption of the Arrhenius definition.

Arrhenius

The Arrhenius definition of acid-base reactions is a more simplified acid-base concept devised by Svante Arrhenius, which was used to provide a modern definition of bases that followed from his work with Friedrich Wilhelm Ostwald in establishing the presence of ions in aqueous solution in 1884, and led to Arrhenius receiving the Nobel Prize in Chemistry in 1903 for "recognition of the extraordinary services rendered to the advancement of chemistry by his electrolytic theory of dissociation".

As defined at the time of discovery, acid-base reactions are characterized by Arrhenius acids, which dissociate in aqueous solution form hydrogen or the later-termed oxonium (H_3O^+) ions, and Arrhenius bases which form hydroxide (OH^-) ions. More recent IUPAC recommendations now suggest the newer term "hydronium" be used in favor of the older accepted term "oxonium" to illustrate reaction mechanisms such as those defined in the Brønsted-Lowry and solvent system definitions more clearly, with the Arrhenius definition serving as a simple general outline of acid-base characterMore succinctly, the Arrhenius definition can be surmised as, "Arrhenius acids form hydrogen ions in aqueous solution with Arrhenius bases forming hydroxide ions."

The *universal aqueous acid-base definition* of the Arrhenius concept is described as the formation of water from hydrogen and hydroxide ions, or hydronium ions and hydroxide ions produced from the dissociation of an acid and base in aqueous solution ($2H_2O \rightarrow OH^- + H_3O^+$), which leads to the definition that in Arrhenius acid-base reactions, a salt and water is formed from the reaction between an acid and a base— in more simple scientific definitions, this form of reaction is called a Neutralization reaction.

$$acid^+ + base^- \rightarrow salt + water$$

The positive ion from a base can form a salt with the negative ion from an acid. For example, two moles of the base sodium hydroxide (NaOH) can combine with one mole of sulfuric acid (H_2SO_4) to form two moles of water and one mole of sodium sulfate.

$$2NaOH + H_2SO_4 \rightarrow 2\,H_2O + Na_2SO_4$$

Brønsted-Lowry Definition

The Brønsted-Lowry definition, formulated independently by its two proponents Johannes Nicolaus Brønsted and Martin Lowry in 1923 is based upon the idea of protonation of bases through the de-protonation of acids—more commonly referred to as the ability of acids to "donate" hydrogen ions (H^+) or protons to bases, which "accept" them. In contrast to the Arrhenius definition, the Brønsted-Lowry definition refers to the products of an acid-base reaction as conjugate acids and bases to refer to the relation of one proton, and to indicate that there has been a reaction between the two quantities, rather than a "formation" of salt and water, as explained in the Arrhenius definition.

It defines that in reactions, there is the donation and reception of a proton, which essentially refers to the removal of a hydrogen ion bonded within a compound and its reaction with another compound, and not the removal of a proton from the nucleus of an atom, which would require inordinate amounts of energy not attainable through the simple dissociation of acids. In differentiation from the Arrhenius definition, the Brønsted-Lowry definition postulates that for each acid, there is a conjugate acid and base or "*conjugate acid-base pair*" that is formed through a complete reaction, which also includes water, which is amphoteric.

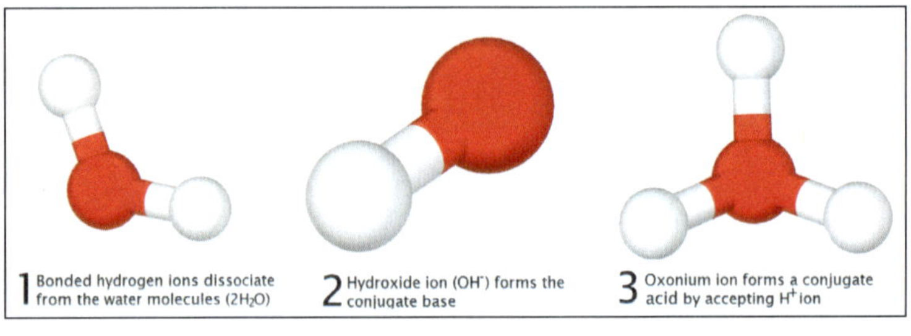

$$AH + B \rightarrow BH^+ + A^-$$

General formula for representing Brønsted-Lowry reactions.

$$HCl\ (aq) + H_2O \rightarrow H_3O^+\ (aq) + Cl^-\ (aq)$$

Hydrochloric acid completely reacts with water to form the hydronium and chloride ions.

$$CH_3COOH + NH_3 \rightarrow NH_4^+ + CH_3COO^-$$

Acetic acid reacts incompletely with ammonia, no hydronium ions being produced.

Lewis

The Lewis definition of acid-base reactions, devised by Gilbert N. Lewis in 1923 is an encompassing theory to the Brønsted-Lowry and solvent-system definitions with regards to the premise of a donation mechanism,which conversely attributes the donation of electron pairs from bases and the acceptance by acids, rather than protons or other bonded substances and spans both aqueous and non-aqueous reactions.

$$Ag^+ + 2{:}\,NH_3 \rightarrow [H_3N{:}Ag{:}NH_3]^+$$

A silver cation reacts as an acid with ammonia which acts as an electron-pair donor, forming an ammonia-silver adduct.

In reactions between Lewis acids and bases, there is the formation of an adduct when the highest occupied molecular orbital (HOMO) of a molecule, such as NH_3 with available lone electron pair(s) donates lone pairs of electrons to the electron-deficient molecule›s lowest unoccupied molecular orbital (LUMO) through a co-ordinate covalent bond; in such a reaction, the HOMO-interacting molecule acts as a base, and the LUMO-interacting molecule acts as an acid. In highly-polar molecules, such as boron trifluoride (BF_3), the most electronegative element pulls electrons towards its own orbitals, providing a more positive charge on the less-electronegative element and a difference in its electronic structure due to the axial or equatorial orbiting positions of its electrons, causing repulsive effects from *lone pair-bonding pair* (Lp-Bp) interactions between bonded atoms in excess of those already provided by *bonding pair-bonding pair* (Bp-Bp) interactions. Adducts involving metal ions are referred to as co-ordination compounds.

Solvent-system

This definition is based on a generalization of the earlier Arrhenius definition to all autodissociating solvents. In all such solvents there is a certain concentration of a positive species, solvonium cations and negative species, solvate anions, in equilibrium with the neutral solvent molecules. For example:

$$2H_2O \rightleftharpoons H_3O^+ \text{ (hydronium)} + OH^- \text{ (hydroxide)}$$

$$2NH_3 \rightleftharpoons NH_4^+ \text{ (ammonium)} + NH_2^- \text{ (amide)}$$

or even some aprotic systems:

$$N_2O_4 \rightleftharpoons NO^+ \text{ (nitrosonium)} + NO_3^- \text{ (nitrate)}$$

$$2SbCl_3 \rightleftharpoons SbCl_2^+ \text{ (dichloroantimonium)} + SbCl_4^- \text{ (tetrachloroantimonate)}$$

A solute causing an increase in the concentration of the solvonium ions and a decrease in the solvate ions is an acid and one causing the reverse is a base. Thus, in liquid ammonia, KNH_2 (supplying NH_2^-) is a strong base, and NH_4NO_3 (supplying NH_4^+) is a strong acid. In liquid sulfur dioxide (SO_2), thionyl compounds (supplying SO^{2+}) behave as acids, and sulfites (supplying SO_3^{2-}) behave as bases.

Here are some nonaqueous acid-base reactions in liquid ammonia:

$$2NaNH_2 \text{ (base)} + Zn(NH_2)_2 \text{ (amphiphilic amide)} \rightarrow Na_2[Zn(NH_2)_4]$$

$$2NH_4I \text{ (acid)} + Zn(NH_2)_2 \text{ (amphiphilic amide)} \rightarrow [Zn(NH_3)_4)]I_2$$

Nitric acid can be a base in liquid sulfuric acid:

$$HNO_3\text{ (base)} + 2H_2SO_4 \rightarrow NO_2^+ + H_3O^+ + 2HSO_4^-$$

And things become even stranger in the aprotic world, for example in liquid N_2O_4:

$$AgNO_3\text{ (base)} + NOCl\text{ (acid)} \rightarrow N_2O_4 + AgCl$$

Since solvent-system definition depends on the solvent as well as on the compound itself, the same compound can change its role depending on the choice of the solvent. Thus, $HClO_4$ is a strong acid in water, a weak acid in acetic acid, and a weak base in fluorosulfonic acid.

Other Acid-base Theories

Usanovich

The most general definition is that of the Russian chemist Mikhail Usanovich, and can basically be summarized as defining an acid as "anything that accepts negative species or donates positive ones," and a base as the reverse. This tends to overlap the concept of redox (oxidation-reduction), and so is not highly favored by chemists. This is because redox reactions focus more on physical electron transfer processes, rather than bond making/bond breaking processes, although the distinction between these two processes is somewhat ambiguous.

Lux-Flood

This definition, proposed by German chemist Hermann Lux in 1939, further improved by Håkon Flood circa 1947 and now commonly used in modern geochemistry and electrochemistry of molten salts, describes an acid as an oxide ion acceptor and a base as an oxide ion donor. For example:

$$MgO\text{ (base)} + CO_2\text{ (acid)} \rightarrow MgCO_3$$

$$CaO\text{ (base)} + SiO_2\text{ (acid)} \rightarrow CaSiO_3$$

$$NO_3^-\text{ (base)} + S_2O_7^{2-}\text{ (acid)} \rightarrow NO_2^+ + 2SO_4^{2-}$$

Pearson

In 1963, Ralph Pearson proposed an advanced qualitative concept known as Hard Soft Acid Base principle, later made quantitative with help of Robert Parr in 1984. ‹Hard› applies to species which are small, have high charge states, and are weakly polarizable. ‹Soft› applies to species which are large, have low charge states and are strongly polarizable. Acids and bases interact and the most stable interactions are hard-hard and soft-soft. This theory has found use in both organic and inorganic chemistry.

Oxidation-reduction Reaction

Oxidation-reduction reaction or redox reaction is any chemical reaction in which the oxidation number of a participating chemical species changes. The term covers a large and diverse body of processes. Many oxidation-reduction reactions are as common and familiar as fire, the rusting and dissolution of metals, the browning of fruit, and respiration and photosynthesis—basic life functions.

Major Classifications

Most oxidation-reduction (redox) processes involve the transfer of oxygen atoms, hydrogen atoms, or electrons, with all three processes sharing two important characteristics: (1) they are coupled—i.e. in any oxidation reaction a reciprocal reduction occurs, and (2) they involve a characteristic net chemical change—i.e. an atom or electron goes from one unit of matter to another. Both reciprocity and net change are illustrated below in examples of the three most common types of oxidation-reduction reactions.

Oxygen-atom Transfer

Carbon reacts with mercury (II) oxide to produce carbon dioxide and mercury metal. This reaction can be written in equation form:

C	+	2HgO	$\rightarrow$	CO_2	+	Hg.
carbon		mercury(II) oxide		carbon dioxide		mercury metal

Carbon, receiving oxygen, is oxidized; mercury(II) oxide, losing oxygen, undergoes the complementary reduction; and the net change is the transfer of two oxygen atoms from mercury(II) oxide units to a carbon atom.

Hydrogen-atom Transfer

Hydrogen atoms are transferred from hydrazine, a compound of nitrogen and hydrogen, to oxygen in the following reaction:

N_2H_4	+	O_2	$\rightarrow$	N_2	+	$2H_2O$
hydrazine		oxygen		nitrogen		water

Hydrazine, losing hydrogen, is oxidized to molecular nitrogen, while oxygen, gaining hydrogen, is reduced to water.

Electron Transfer

Zinc metal and copper(II) ion react in water solution, producing copper metal and an aqueous (denoted by aq) zinc ion according to the equation,

Zn	+	Cu^{2+}(aq)	$\rightarrow$	Zn^{2+}(aq)	+	Cu
zinc metal		aqueous copper(II) ion		aqueous zinc(II) ion		copper metal

With the transfer of two of its electrons, the zinc metal is oxidized, becoming an aqueous zinc ion, while the copper(II) ion, gaining electrons, is reduced to copper metal. Net change is the transfer of two electrons, lost by zinc and acquired by copper.

Because of their complementary nature, the oxidation and reduction processes together are referred to as redox reactions. The reactant that brings about the oxidation is called the oxidizing agent, and that reagent is itself reduced by the reducing agent. In the examples given above, mercury(II) oxide, oxygen, and the copper(II) ion are oxidizing agents, and carbon, hydrazine, and zinc are the reducing agents.

General Theory

Stoichiometric Basis

Describing the redox processes as above conveys no information about the mechanism by which change takes place. A complete description of the net chemical change for a process is known as the stoichiometry of the reaction, which provides the characteristic combining proportions of elements and compounds. Reactions are classified as redox and non-redox on the basis of stoichiometry; oxygen-atom, hydrogen-atom, and electrontransfer are stoichiometric categories.

Oxidation-state Change

Comprehensive definitions of oxidation and reduction have been made possible by modern molecular structure theory. Every atom consists of a positive nucleus, surrounded by negative electrons, which determine the bonding characteristics of each element. In forming chemical bonds, atoms donate, acquire, or share electrons. This makes it possible to assign every atom an oxidation number, which specifies the number of its electrons that can be involved in forming bonds with other atoms. From the particular atoms in a molecule and their known bonding capacities, the bonding pattern within a molecule is determined, and each atom is regarded as being in a specific oxidation state, expressed by an oxidation number.

Redox processes are defined as reactions accompanied by oxidation-state changes: an increase in an atom's oxidation number corresponds to an oxidation; a decrease, to a reduction. In this generalized theory, three examples of ways in which oxidation-state changes can occur are by oxygen-atom (gain, oxidation; loss, reduction), hydrogen-atom (loss, oxidation; gain, reduction), and electron (loss, oxidation; gain, reduction) transfer. The oxidation-state change definition is usually compatible with the above rules for applying the oxygen-atom-transfer and hydrogen-atom-transfer criteria and always compatible with the electron-transfer criterionwhen it is applicable. The oxidation state of any atom is indicated by a roman numeral following the name or symbol for the element. Thus, iron(III), or Fe(III), means iron in an oxidation state of +3. The uncombined Fe(III) ion is simply Fe^{3+}.

Molecular oxygen is a conspicuously important oxidizing agent. It will directly oxidize all but a few of the metals and most of the nonmetals as well. Often these direct oxidations lead to normal oxides, such as those of lithium (Li), zinc (Zn), phosphorus (P), and sulfur (S).

$$4Li + O_2 \rightarrow 2Li_2O(\text{lithium oxide})$$

$$2Zn + O_2 \rightarrow 2ZnO(\text{zinc oxide})$$

$$4P + 5O_2 \rightarrow P_4O_{10}(\text{phosphorus pentoxide})$$

$$S + O_2 \rightarrow SO_2(\text{sulfur dioxide})$$

Organic foodstuffs are oxidized to carbon dioxide and water in respiration. The reaction stoichiometry can be illustrated for glucose, a simple sugar:

$C_6H_{12}O_6$	+	$6O_2$	$\rightarrow$	$6CO_2$	+	$6H_2O$.
glucose		oxygen		carbon dioxide		water

Although the oxygen-glucose reaction is slow at ambient temperatures outside the living cell, it proceeds quickly under the influence of enzymatic catalysis within the body. Essentially all organic compounds react with oxygen under appropriate conditions, but the reaction rates at ordinary temperatures and pressures vary greatly.

Many other oxidizing agents serve as oxygen-atom. Hydrogen peroxide (H_2O_2), acid chromate ion ($HCrO_4^-$), and hypochlorous acid (HClO) are reagents often used in oxygen-atom-transfer reactions—for example, in the following reactions:

H_2O_2	+	Cl^-	$\rightarrow$	OCl^-	+	H_2O
hydrogen peroxide		chloride ion		hypochlorite ion		water

$2HCrO_4^-$	+	$3H_3C\overset{O}{\overset{\|\|}{C}}H$	+	$8H^+$	$\rightarrow$
acid chromate ion		acetaldehyde		hydrogen ion	

$3H_3C\overset{O}{\overset{\|\|}{C}}OH$	+	$2Cr^{3+}(aq)$	+	$5H_2O$
acetic acid		chromium(III) ion		water

$5HClO$	+	I_2	+	H_2O	$\rightarrow$	$5HCl$	+	$2HIO_3$
hypochlorous acid		iodine		water		hydrogen chloride		iodic acid

In the simplest hydrogen-atom transfers, molecular hydrogen serves as the hydrogen-atom. The hydrogenations of ethylene and of molecular nitrogen are illustrative in the following equations,

$H_2C{=}CH_2$	+	H_2	$\rightarrow$	H_3CCH_3
ethylene		hydrogen		ethane

N_2	+	$3H_2$	$\rightarrow$	$2NH_3$
nitrogen		hydrogen		ammonia

Reactions of molecular hydrogen are characteristically slow at ordinary temperatures. The hydrogenation of molecular nitrogen and of olefins such as ethylene (an olefin is an unsaturated hydrocarbon compound; it has at least two adjacent carbon atoms joined by a double bond to which other atoms or groups of atoms can be joined directly) is a process of extraordinary commercial importance and requires catalysts to occur at useful rates.

Hydrogen-atom transfer from an organic molecule to a suitable acceptor is a common mode of organic oxidation. The oxidation of formic acid by permanganate and that of ethanol by acid chromate share stoichiometry that features hydrogen-atom loss by the organic species, as shown in the following equations,

$3H\overset{O}{\overset{\|}{C}}OH$	+	$2MnO_4^-$	+	$2H^+$	$\rightarrow$
formic acid		permanganate ion		hydrogen ion	
$3CO_2$	+	$2MnO_2$	+	$4H_2O$	
carbon dioxide		manganese dioxide		water	
$3CH_3CH_2OH$	+	$2HCrO_4^-$	+	$8H^+$	$\rightarrow$
ethanol		acid chromate ion		hydrogen ion	
$8H_2O$	+	$3CH_3CHO$	+	$2Cr^{3+}(aq)$	
water		acetaldehyde		aqueous chromium(III) ion	

The oxidizing agents permanganate and acid chromate, typical of many hydrogen-atom acceptors, undergo complicated changes rather than simple hydrogen-atom addition.

Electron-transfer stoichiometry is usually associated with metal ions in aqueous solution, as shown in the following equations,

$Fe^{2+}(aq)$	+	$Co^{3+}(aq)$	$\rightarrow$	$Fe^{3+}(aq)$	+	$Co^{2+}(aq)$
iron(II) ion		cobalt(III) ion		iron(III) ion		cobalt(II) ion
$Tl^{+}(aq)$	+	$2Ce^{4+}(aq)$	$\rightarrow$	$Tl^{3+}(aq)$	+	$2Ce^{3+}(aq)$
thallium(I) ion		cerium(IV) ion		thallium(III) ion		cerium(III) ion

Many positively charged metal ions have been shown to be bonded to water molecules, so that their electron-transfer reaction occurs between rather complex molecular groups. The iron ion formulas above, for example, are more properly written as $[Fe(H_2O)_6]^{2+}$ and $[Fe(H_2O)_6]^{3+}$ to reflect the presence of six water molecules bonded to the metal ion. Simple electron transfer between free ions is known only in the gas phase, as in this argon-sodiumreaction,

$Ar^{+}(g)$	+	$Na(g)$	$\rightarrow$	$Na^{+}(g)$	+	$Ar(g)$
argon ion		sodium		sodium ion		argon

Several other types of redox reactions do not fall in the oxygen-atom, hydrogen-atom, or electron-transfer categories. Among these are reactions of fluorine, chlorine, bromine, and iodine. These four elements, known as the halogens, form diatomic molecules, which are versatile oxidizing agents. The following examples are typical,

$3F_2$ fluorine	+	S sulfur	$\rightarrow$	SF_6 sulfur hexafluoride		
Cl_2 chlorine	+	$2Fe^{2+}(aq)$ iron(II) ion	$\rightarrow$	$2Cl^-$ chloride ion	+	$2Fe^{3+}(aq)$ iron(III) ion
Br_2 bromine	+	$H_2C=CH_2$ ethylene	$\rightarrow$	$BrCH_2CH_2Br$ ethylene dibromide		
I_2 iodine	+	$2S_2O_3^{2-}$ thiosulfate ion	$\rightarrow$	$2I^-$ iodide ion	+	$S_4O_6^{2-}$ polythionate ion

Such reactions often qualify as redox processes only in the broad sense that oxidation-state changes occur. The oxidation-state characterization extends oxidation-reduction chemistry to include examples from the reactions of all the chemical elements.

Significance of Redox Reactions

Oxidation-reduction reactions have vast importance not only in chemistry but in geology and biology as well. The surface of Earth is a redox boundary between the planet's reduced metallic core and an oxidizing atmosphere. Earth's crust is largely composed of metal oxides, and the oceans are filled with water, an oxide of hydrogen. The tendency of nearly all surface materials to be oxidized by the atmosphere is reversed by the life process of photosynthesis. Because they are constantly renewed by the photosynthetic reduction of carbon dioxide, life's complex compounds can continue to exist on Earth's surface.

For similar reasons, much of chemical technology hinges on the reduction of materials to oxidation states lower than those that occur in nature. Such basic chemical products as ammonia, hydrogen, and nearly all the metals are produced by reductive industrial processes. When not used as structural materials, these products are reoxidized in their commercial applications. The weathering of materials, including wood, metals, and plastics, is oxidative, since, as the products of technological or photosynthetic reductions, they are in oxidation states lower than those stable in the atmosphere.

Solar radiation is converted to useful energy by a redox cycle that operates continually on a global scale. Photosynthesis converts radiant energy into chemical potential energy by reducing carbon compounds to low oxidation states, and this chemical energy is recovered either through enzymatic oxidations at ambient temperatures or during combustion at elevated temperatures.

Oxidation States

The idea of assigning an oxidation state to each of the atoms in a molecule evolved from the electron-pair concept of the chemical bond. Atoms within a molecule are held together by the force of attraction that the nuclei of two or more of them exert on electrons in the space between them. In many cases this sharing of electrons can be regarded as involving electron-pair bonds between adjacent nuclei. Electron-pair bonding is often diagrammed so as to show all the bonding and nonbonding valence electrons—e.g., the structures of atomic hydrogen, atomic chlorine, and hydrogen chloride shown below (each dot represents one valence electron):

H.	+	:C̈l. (with two dots below Cl)	→	H :C̈l: (with two dots below Cl)
hydrogen atom		chlorine atom		hydrogen chloride molecule

The hydrogen chloride diagram reflects the presence, in the internuclear region, of two electrons that are under the mutual attractive influence of both the hydrogen and chlorine nuclei. Oxidation states for the hydrogen and chlorine in HCl are assigned according to the net charges that remain on H and Cl when the shared electrons are assigned to the atom that has the greater attraction for them. Through physical measurements on isolated atoms and simple molecules, these relative attractive powers have been determined.

The table lists the electronegativity values for some important elements.

Pauling electronegativities of selected elements	
fluorine	3.98
oxygen	3.44
chlorine	3.16
nitrogen	3.04
bromine	2.96
iodine	2.66
sulfur	2.58
carbon	2.55
hydrogen	2.2
phosphorus	2.19
iron	1.83
sodium	0.93

In the hydrogen chloride molecule the chlorine is more electronegative than hydrogen and is, therefore, assigned both shared electrons. Chlorine has seven valence electrons in its neutral state. Having acquired an eighth electron in its reaction with hydrogen, chlorine is considered to have an oxidation state of −1. Hydrogen, on the other hand, is assigned +1, having lost the single valence electron that it has in its neutral state. Charges arrived at in this way are the basis for oxidation-state assignments, conventionally represented by roman numerals, such as in H(I) and Cl(−I) for the constituents of HCl. Because determination of oxidation states is simply a method of conceptually distributing shared electrons to individual atoms, the same number of electrons must be accounted for, before and after such assignment.

Oxygen-atom Transfer Reactions

The oxidation-state concept clarifies the relationship between oxygen-atom, hydrogen-atom, and electron transfer. The oxygen- and hydrogen-transfer criteria apply only when oxygen and hydrogen occur in their typical oxidation states. An example of an appropriate reaction involving oxygen-atom transfer is the reduction of ferrous oxide by carbon monoxide,

$$\underset{\text{(ferrous oxide)}}{\underset{\text{iron(II) oxide}}{FeO}} + \underset{\text{monoxide}}{\underset{\text{carbon}}{CO}} \rightarrow \underset{\text{dioxide}}{\underset{\text{carbon}}{CO_2}} + \underset{\text{iron}}{Fe}$$

In terms of oxidation-state changes, this oxygen-atom transfer is equivalent to the two-electron reduction of iron and complementary two-electron oxidation of carbon:

$$Fe(II) + C(II) \rightarrow C(IV) + Fe(0).$$

Oxygen, which occurs in the oxidation state −2 in both reactants and products in the first equation, is not shown in the second. In transferring, the oxygen atom leaves two electrons behind, causing the reduction of iron, and acquires two electrons from the carbon atom, oxidizing the carbon.

In a similar way, the hydrogenation of ethylene corresponds to a two-electron reduction of the two-carbon skeleton:

$$\underset{\text{hydrogen}}{H_2} + \underset{\text{ethylene}}{H_2C{=}CH_2} \rightarrow \underset{\text{ethane}}{H_3CCH_3}$$

$$\underset{\text{oxidation state zero}}{\underset{\text{hydrogen in}}{2\,H(0)}} + \underset{\text{oxidation state-2}}{\underset{\text{carbon in}}{2\,C(-II)}} \rightarrow$$

$$\underset{\text{oxidation state+1}}{\underset{\text{hydrogen in}}{2\,H(I)}} + \underset{\text{oxidation state-3}}{\underset{\text{carbon in}}{2\,C(-III).}}$$

In this example also, the second equation includes only the atoms that change oxidation states: the four hydrogen atoms initially present in ethylene are in the +1 oxidation state in both reactants and products and are therefore omitted. Each of the two neutral hydrogen atoms can be regarded as giving up an electron to, and thereby reducing, one of the carbon atoms. This example also demonstrates clearly that the oxidation that complements the reduction of ethylene is that of the two hydrogen atoms in H_2—namely, from the 0 to the +1 oxidation state. General application of the oxidation-state concept leads to a formal viewpoint toward all redox reactions as electron-transfer reactions.

Half Reactions

One of the basic reasons that the concept of oxidation-reduction reactions helps to correlate chemical knowledge is that a particular oxidation or reduction can often be carried out by a wide variety of oxidizing or reducing agents.

Reduction of the iron(III) ion to the iron(II) ion by four different reducing agents provides an example:

$Fe^{3+}(aq)$ +	$V^{2+}(aq)$	→ $Fe^{2+}(aq)$ +	$V^{3+}(aq)$
iron(III) ion	vanadium(II) ion	iron(II) ion	vanadium(III) ion

$2\,Fe^{3+}(aq)$ +	$Zn(s)$	→ $2\,fe^{2+}(aq)$ +	$Zn^{2+}(aq)$
iron(III) ion	zinc metal	iron(II) ion	zinc(II) ion

$2\,Fe^{3+}(aq)$ +	$2\,I^{-}$	→ $2\,Fe^{2+}(aq)$ +	I_2
iron(III) ion	iodide ion	iron(II) ion	iodine

$2\,Fe^{3+}(aq)$ +	$C_6H_4(OH)_2$	→ $2\,Fe^{2+}(aq)$ +	$C_6H_4O_2$ +	$2\,H^{+}$
iron(III) ion	hydroquinone	iron(II) ion	quinone	hydrogen ion

Production of the same change in the aqueous iron(III) ion by different reductants emphasizes the fact that the reduction is a characteristic reaction of the iron system itself, and, therefore, the process may be written without specifying the identity of the reducing agent in the following way:

$Fe^{3+}(aq)$ +	e^{-}	→ $Fe^{2+}(aq)$.
iron(III) ion	electron	iron(II) ion

Hypothetical equations of this type are known as half reactions. The symbol e–, which stands for an electron, serves as a reminder that an unspecified reducing agent is required to bring about the change. Half reactions can be written, equally, for the reducing agents in the four reactions with ferric ion:

$V^{2+}(aq)$	→ $V^{3+}(aq)$	+ e^{-}
vanadium(II) ion	vanadium(III) ion	electron

$Zn(s)$	→ $Zn^{2+}(aq)$ +	$2\,e^{-}$
zinc(0) metal	zinc(II) ion	electron

$2\,I^{-}$	→ I_2 +	$2\,e^{-}$
iodide ion	iodine	electron

$C_6H_4(OH)_2$	→ $C_6H_4O_2$ +	$2\,e^{-}$ +	$2\,H^{+}$
hydroquinone	quinone	electron	hydrogenion

Although hypothetical, half reactions are properly balanced chemical processes. Since $V^{2+}(aq)$ increases its oxidation number by one, from +2 to +3, in the first half reaction, an electron is shown as a product of the change. Similarly, two electrons are produced when the oxidation number of zinc increases from 0 to +2 in the second half reaction. When half reactions for hypothetical isolated oxidations and reductions are combined, the electrons must cancel if the equation for a possible overall chemical reaction is to result.

The use of half reactions is a natural outgrowth of the application of the electron-transfer concept to redox reactions. Since the oxidation-state principle allows any redox reaction to be analyzed in terms of electron transfer, it follows that all redox reactions can be broken down into a complementary pair of hypothetical half reactions. Electrochemical cells (in which chemical energy can be converted to electrical energy, and vice versa) provide some physical reality to the half-reaction idea. Oxidation and reduction half reactions can be carried out in separate compartments of electrochemical cells, with the electrons flowing through a connecting wire and the circuit completed by some arrangement for ion migration between the two compartments (but the migration need not involve any of the materials of the oxidation-reduction reactions themselves).

Redox Potentials for Common Half Reactions

The analysis of the electrical potential, or voltage, developed by pairing various half reactions in electrochemical cells has led to the determination of redox potentials for a substantial number of common half reactions. While a detailed description of redox potentials requires the methods of thermodynamics (the branch of physics concerned with the role played by heat in the transformation of matter or energy), a great deal of useful information can be obtained from redox potentials with minimal recourse to formal theory. Basically, a table of half-cell potentials is a summary of the relative tendencies of different oxidations and reductions to occur.

The table of standard reduction potentials lists selected half reactions and their corresponding reduction potentials (which are symbolized by $E°$). The physical significance of the values is directly linked to several agreements about their use. First, the greater the value of $E°$ (the reduction potential), the greater the tendency of a half reaction to proceed from left to right (as written). The half reactions in the table are listed from top to bottom in order of decreasing $E°$: the higher a reaction's position on the list, the greater the tendency of the reactants to accept electrons. In other words, reagents high on the list, such as fluorine gas (F_2) and permanganate ion (MnO_4^-) are strong oxidizing agents. Second, the reduction of hydrogen ions (H^+) to hydrogen gas (H_2) is arbitrarily assigned the value 0 volts. Half cells with positive reduction potentials involve reactants that are more readily reduced than H^+; conversely, those with negative potentials involve reactants that are more difficult to reduce than hydrogen ions.

With the aid of reduction potentials, it is possible to predict whether a particular oxidation-reduction reaction can occur. The predictions require breaking down the overall reaction into two half reactions of known reduction potentials. For example, if a strip of zinc metal is dipped into a solution containing copper(II) ion, the possibility exists for a redox process, which can be regarded as the sum of the half reactions aqueous zinc ion (Zn^{2+}[aq]) to zinc metal (Zn[s]) and aqueous copper ion (Cu^{2+}[aq]) to copper metal (Cu[s]), as follows,

$$\underset{\text{zinc(0) metal}}{Zn(s)} + \underset{\text{copper ion}}{Cu^{2+}(aq)} \rightarrow \underset{\text{zinc(II) ion}}{Zn^{2+}(aq)} + \underset{\text{copper(0) metal}}{Cu(s)}$$

$$Cu^{2+}(aq) + 2e^- \longrightarrow Cu(s) \qquad +0.34 \text{ volt}$$

$$Zn(s) \longrightarrow Zn^{2+}(aq) + 2e^- \qquad +0.76$$

$$\underset{\text{copper(II) ion}}{Cu^{2+}(aq)} + \underset{\text{zinc metal}}{Zn(s)} \xrightarrow{\text{spontaneous}} \underset{\text{copper metal}}{Cu(s)} + \underset{\text{zinc(II) ion}}{Zn^{2+}(aq)} \qquad +1.10$$

The resulting *E*° value for the net reaction, +1.10 volts, measures the tendency of the net reaction to occur. If *E*° for a particular net reaction is positive, the process may be expected to occur spontaneously when the reactants are mixed at specified concentrations (one mole per litre; Oxidation-reduction equilibria). Therefore, it is predicted that copper metal should be deposited on a strip of zinc metal when the latter is immersed in a solution of a copper(II) salt. This reaction is, in fact, readily observed in the laboratory. A more specific physical interpretation of the +1.10 volt value is that it represents the voltage that would be produced by an ideal electrochemical cell based on the copper(II) ion to copper metal and zinc(II) ion to zinc metal half reactions with all the reagents at specified concentrations.

When the same two half cells are combined, with both their directions (and therefore the signs of their redox potentials) reversed, it is predicted that the reverse reaction, the depositing of zinc metal from a zinc(II) ion solution onto a copper strip, will not occur spontaneously. As in the case of *E*° values for half reactions, those for net redox reactions also change sign when the direction of the reaction is reversed.

$$\begin{array}{lr} Cu(s) \rightarrow Cu^{2+}(aq)+2e^{-} & -0.34 \text{ volt} \\ Zn^{2+}(aq)+2e^{-} \rightarrow Zn(s) & -0.76 \\ \hline Zn^{2+}(aq)+Cu(s) \xrightarrow{\text{nonspontaneous}} Cu^{2+}(aq)+Zn(s) & -1.10 \end{array}$$

The results of the copper-zinc system can be applied more generally to the half reactions in the table of standard reduction potentials. For example, copper(II) ion in water (Cu^{2+}[aq]) is an oxidant strong enough to force a half reaction lower on the table to proceed spontaneously in the opposite direction of that written. Therefore, not only is copper(II) ion expected to oxidize zinc metal (Zn[s]) to zinc(II) ion (Zn^{2+}[aq]); it is also predicted to oxidize hydrogen gas (H_2[g]) to hydrogen ion (H^+) and sodium metal (Na[s]) to sodium ion (Na^+).

Similarly, fluorine gas (F_2[g]), the strongest oxidant listed in the table of standard reduction potentials, is predicted to oxidize spontaneously the products of all the other half reactions in the table. In contrast, the strongest reducing agent is solid sodium metal (Na[s]), and it is expected spontaneously to reduce the reactants of all the other half cells.

Selected values of standard reduction potentials are given in the table.

Selected values of standard reduction potentials									
Half reactions									E° (volts)
F_2(g) fluorine(0)	+	$2e^-$ electrons	→	$2F^-$ fluoride (−I) ion					2.87
MnO_4^- permanganate ion	+	$8H^+$ hydrogen (I) ions	+	$5e^-$ electrons	→	Mn^{2+}(aq) manganese(II) ion	+	$4H_2O$ water	1.51
Cl_2(g) chlorine(0)	+	$2e^-$ electrons	→	$2Cl^-$ chloride (−I) ions					1.36
O_2(g) oxygen(0)	+	$4H^+$ hydrogen (I) ions	+	$4e^-$ electrons	→	$2H_2O$ water			1.23
Fe^{3+}(aq) iron(III) ion	+	e^- electron	→	Fe^{2+}(aq) iron (II) ion					0.77
Cu^{2+}(aq) copper(II) ion	+	$2e^-$ electrons	→	Cu(s) copper(0)					0.34

$2H^+$ hydrogen(I) ions	+	$2e^-$ electrons	→	$H_2(g)$ hydrogen(0)					0.00
$Zn^{2+}(aq)$ zinc(II) ion	+	$2e^-$ electrons	→	Zn(s) zinc(0)					−0.76
Na^+ sodium(I) ion	+	e^- electron	→	Na(s) sodium(0)					−2.71
The identifications in parentheses refer to the physical state of the substance: (g), gas; (aq), hydrated positive ion in water; (s), pure solid.									

Oxidation-reduction Equilibria

In practice many chemical reactions can be carried out in either direction, depending on the conditions. The spontaneous direction predicted for a particular redox reaction by half-cell potentials is appropriate to a standard set of reaction conditions. Specifically, the temperature is assumed to be 25 °C with reagents at specified concentrations. Gases are present at one atmosphere pressure and solutes at one mole per litre (one molecular weight in grams dissolved in one litre of solution) concentration (1M). Solids are assumed to be in contact with the reaction solution in their normal stable forms, and water is always taken to be present as the solvent. Many practical problems can be solved directly with standard reduction potentials.

The usefulness of reduction potentials is greatly extended, however, by a thermodynamic relationship known as the Nernst equation, which makes it possible to calculate changes in half-cell potentials that will be produced by deviations from standard concentration conditions. In the reaction between zinc metal and copper(II) ion, standard conditions for zinc and copper metal require simply that both solids be present in contact with the solution; the $E°$ values are not affected by either the total or proportionate amounts of the two metals. The calculation that the overall reaction is spontaneous by +1.10 volts is based on standard one mole per litre (1M) concentrations for aqueous zinc(II) ion (Zn^{2+}[aq]) and aqueous copper(II) ion (Cu^{2+}[aq]). Using the Nernst equation it is found that $E°$ for the overall reaction will be +1.10 volts as long as both ions are present in equal concentrations, regardless of the concentration level.

On the other hand, if the ratio of the zinc(II) to copper(II) ion concentrations is increased, the reduction potential ($E°$) falls until, at a very high preponderance of zinc ion, $E°$ becomes 0 volt. At this point, there is no net tendency for the reaction to proceed spontaneously in either direction. If the zinc(II) to copper(II) ion ratio is increased further, the direction of spontaneity reverses, and zinc ion spontaneously oxidizes copper metal. In practice, such high zinc(II) to copper(II) ion concentration ratios are unattainable, which means that the reaction can only be carried out spontaneously with copper(II) ion oxidizing zinc metal. Many reactions with $E°$ values smaller than +1.10 volts under standard conditions can be carried out in either direction by adjusting the ratio of product and reactant concentrations. The point at which $E° = 0$ volt represents a state of chemical equilibrium. When chemical reactions are at equilibrium, the concentrations of the reagents do not change with time, since net reaction is not spontaneous in either direction. Measurements of half-cell potentials combined with Nernst-equation calculations are a powerful technique for determining the concentration conditions that correspond to chemical equilibrium.

Reaction Rates

Predictability

There are practical limitations on predictions of the direction of spontaneity for a chemical reaction, the most important arising from the problem of reaction rates. An analogy can be made with the simple physical system of a block on a sloping plane. Because of the favourable energy change, the block tends spontaneously to slide down, rather than up, the slope, and, at mechanical equilibrium, it will be at the bottom of the slope, since that is the position of lowest gravitational energy. How rapidly the block slides down is a more complex question, since it depends on the amount and kind of friction present. The direction of spontaneity for a chemical reaction is analogous to the downhill direction for a sliding block, and chemical equilibrium is analogous to the position at the bottom of the slope; the rate at which equilibrium is approached depends on the efficiency of the available reaction processes. Between zinc metal and aqueous copper(II) ion, the reaction proceeds without observable delay, but various other spontaneous redox processes proceed at imperceptibly slow rates under ordinary conditions.

Biological Processes

A particularly significant illustration of the role of mechanisms in determining the rates of redox reactions concerns respiration, the central energy-producing process of life. Foodstuffs that are oxidized by molecular oxygen during respiration are quite unreactive with oxygen before ingestion. Such high-energy foods as grains and sugar can resist the atmosphere indefinitely but are rapidly converted to carbon dioxide and water through combination with oxygen during respiratory metabolism. The situation is exemplified by the behaviour of glucose at ambient temperatures.

The significance of the different rate behaviour of high-energy foods inside and outside the cell has been dramatized by Albert Szent-Györgyi, a Hungarian-born American biochemist and a pioneering researcher in the chemical mechanism of respiration.

Living systems are able to use respiratory oxidation as an energy only because the same reactions are slow outside the cell. In return for providing an efficient mechanism for the oxidation of foods, the cell gains control over the disposition of the liberated chemical energy.

Examples such as the chemistry of respiration make clear the importance of determining the rates and mechanisms of redox reactions. Often questions are difficult to answer even in regard to relatively simple reactions. It has been pointed out that many redox processes can be categorized as oxygen-atom-, hydrogen-atom-, or electron-transfer processes. These categories describe the net changes that are involved but provide no insight into the mechanisms of the reactions.

Mechanisms of Redox Reactions

Some of the problems associated with formulating descriptions of the mechanisms are illustrated by the reaction between two metal ions that undergo complementary, one-unit changes in oxidation state,

$$M(\text{III}) + N(\text{II}) \rightarrow M(\text{II}) + N(\text{III}).$$

There are many different metal ions, designated with the letters M and N, which participate in redox reactions with this basic stoichiometry. To define the possible mechanisms with any precision, it is necessary to identify the groups that are directly bonded to the metals, as well as to specify which particular metals take part in the reactions. One relatively simple example is the chromium(II)-iron(III) reaction. Both chromium and iron aquo ions are surrounded by six water molecules in both the +2 and +3 oxidation states.

$$\underset{\substack{\text{hexaaquoiron (III)}\\ \text{ion}}}{Fe(H_2O)_6^{3+}} + \underset{\substack{\text{hexaaquochromium(II)}\\ \text{ion}}}{Cr(H_2O)_6^{2+}} \rightarrow$$

$$\underset{\substack{\text{hexaaquoiron(II)}\\ \text{ion}}}{Fe(H_2O)_6^{2+}} + \underset{\substack{\text{hexaaquochromium(III)}\\ \text{ion}}}{Cr(H_2O)_6^{3+}}$$

The reaction appears from its stoichiometry to entail simple electron transfer from Cr(II) to Fe(III). There are, however, other possibilities. Their variety is most evident if it is assumed that the oxidizing agent, hexaaquoiron(III) ($Fe[H_2O]_6^{3+}$), dissociates a hydrogen ion before reacting with hexaaquochromium(II) ($Cr[H_2O]_6^{2+}$). The hexaaquoiron(III) ion is then converted to a hydroxopentaaquoiron(III) ion, known to be an important reactant in the Fe(III)-Cr(II) reaction. Its formation requires no change in the oxidation state of iron and should be regarded as a simple acid-base reaction. With hydroxopentaaquoiron(III) ion, $(H_2O)_5FeOH^{2+}$, as the oxidizing agent, simple electron transfer does provide one plausible means for the reaction. All the steps in the mechanism must add up to give the overall reaction. Intermediate species such as hydroxopentaaquoiron(III) must cancel out in the addition process by occurring as products in one step and reactants in another.

$$\underset{\substack{\text{hexaaquoiron(III)}\\ \text{ion}}}{Fe(H_2O)_6^{3+}} \rightarrow \underset{\substack{\text{hydroxopenta-}\\ \text{aquoiron(III)}\\ \text{ion}}}{(H_2O)_5\,FeOH^{2+}} + \underset{\substack{\text{hydrogen}\\ \text{ion}}}{H^+}$$

$$\underset{\substack{\text{hydroxopenta-}\\ \text{aquoiron(III)}\\ \text{ion}}}{(H_2O)_5\,FeOH^{2+}} + \underset{\substack{\text{hexaaquo-}\\ \text{chromium(III)}\\ \text{ion}}}{Cr(H_2O)_6^{2+}}$$

$$\xrightarrow[\text{transfer}]{\text{electron}} \underset{\substack{\text{hydroxopenta-}\\ \text{aquoiron(II)}\\ \text{ion}}}{(H_2O)5\,FeOH^+} + \underset{\substack{\text{hexaaquo-}\\ \text{chromium(III)}\\ \text{ion}}}{Cr(H_2O)_6^{3+}}$$

$$\underset{\substack{\text{hydroxopenta-}\\ \text{aquoiron(II)}\\ \text{ion}}}{(H_2O)_5\,FeOH^+} + \underset{\substack{\text{hydrogen}\\ \text{ion}}}{H^+} \rightarrow \underset{\substack{\text{hexaaquoiron}\\ \text{ion}}}{Fe(H_2O)_6^{2+}}$$

$$Fe(H_2O)_6^{3+} + Cr(H_2O)_6^{2+} \rightarrow Fe(H_2O)_6^{2+} + Cr(H_2O)_6^{3+}$$

An alternative to simple electron transfer is the transfer of a hydrogen atom from a water molecule bound to chromium(II) to the hydroxide ion (OH^-) bound to iron(III). A third possibility is the transfer of a neutral OH group (called a hydroxyl radical) from iron(III) to chromium(II). Oxygen and hydrogen occur as oxygen(–II) and hydrogen(I) in both reactants and products. These oxidation states correspond to those present in the hydroxide ion OH^-. Therefore, when an OH group transfers, it must leave an electron behind, reducing the iron, and then reaccept an electron from chromium, oxidizing it. Since both the aqueous iron and chromium ions have a strong tendency to remain surrounded by six oxygen molecules, the most likely method of OH transfer would be the sharing of the transferred oxygen between the two metals at an intermediate stage of the reaction. But iron(III) is not a strong enough oxidizing agent to produce a significant concentration of neutral OH groups from OH^- ions. A more acceptable description of the OH transfer mechanism, therefore, is to regard OH^- as a bridging group through which electron transfer is accomplished as the two metal ion groups share a bonded oxygen atom.

$$[(H_2O)_5 \overset{3+}{Fe}-\underset{|\atop H}{O}-\overset{2+}{Cr}(H_2O)_5]^{4+} \rightarrow [(H_2O)_5 \overset{2+}{Fe}-\underset{|\atop H}{O}-\overset{3+}{Cr}(H_2O)_5]^{4+}$$

Experimentally, it is difficult to distinguish between the alternative mechanisms posed above. Classification schemes for reaction mechanisms are of little value unless they offer alternatives that can be experimentally verified. For that reason the most successful system for classifying the mechanisms of redox reactions between metal ions has been the inner sphere–outer sphere dichotomy. Outer-sphere reactions are those that take place without breaking any bonds between a metal and a group such as water or hydroxide ion bound to it. Both the simple electron- and hydrogen-atom-transfer mechanisms are outer-sphere processes. If, on the other hand, the oxidation-state changes take place within an intermediate product in which the two metals are directly bonded to a common bridging group (such as shown in the equation above), the mechanism is inner-sphere. Formation of the intermediate requires loss of a bound water molecule by one of the metals. The inner sphere–outer sphere distinction can be tested experimentally in some redox systems.

Combustion Reactions

Combustion is a chemical reaction between substances, usually including oxygen and usually accompanied by the generation of heat and light in the form of flame. The rate or speed at which the reactants combine is high, in part because of the nature of the chemical reaction itself and in part because more energy is generated than can escape into the surrounding medium, with the result that the temperature of the reactants is raised to accelerate the reaction even more. A familiar example is a lighted match. When a match is struck, friction heats the head to a temperature at which the chemicals react and generate more heat than can escape into the air, and they burn with a flame. If a wind blows away the heat or the chemicals are moist and friction does not raise the temperature sufficiently, the match goes out. Properly ignited, the heat from the flame raises the temperature of a nearby layer of the matchstick and of oxygen in the air adjacent to it, and the wood and oxygen react in a combustion reaction. When equilibrium between the total heat energies of the reactants and the total heat energies of the products (including the actual heat and light emitted) is reached, combustion stops. Flames have a definable composition and a complex structure; they are said to be multiform and are capable of existing at quite low temperatures, as well as at extremely high temperatures. The emission of light in the flame results from the presence of excited particles and, usually, of charged atoms and molecules and of electrons.

Combustion encompasses a great variety of phenomena with wide application in industry, the sciences, professions, and the home, and the application is based on knowledge of physics, chemistry, and mechanics; their interrelationship becomes particularly evident in treating flame propagation.

In general terms, combustion is one of the most important of chemical reactions and may be considered a culminating step in the oxidation of certain kinds of substances. Though oxidation was once considered to be simply the combination of oxygen with any compound or element, the meaning of the word has been expanded to include any reaction in which atoms lose electrons, thereby becoming oxidized. As has been pointed out, in any oxidation process the oxidizer takes electrons from the oxidizable substance, thereby itself becoming reduced (gaining electrons). Any substance at all can be an oxidizing agent. But these definitions, clear enough when applied to atomic structure to explain chemical reactions, are not as clearly applicable to combustion, which remains, generally speaking, a type of chemical reaction involving oxygen as the oxidizing agent but complicated by the fact that the process includes other kinds of reactions as well and by the fact that it proceeds at an unusually fast pace. Furthermore, most flames have a section in their structure in which, instead of oxidations, reduction reactions occur. Nevertheless, the main event in combustion is often the combining of combustible material with oxygen.

Physical and Chemical Aspects of Combustion

The Chemical Reactions

Combustion, with rare exceptions, is a complex chemical process involving many steps that depend on the properties of the combustible substance. It is initiated by external factors such as heat, light, and sparks. The reaction sets in as the mixture of combustibles attains the ignition temperature. The combustion spreads from the ignition source to the adjacent layer of gas mixture; in turn, each point of the burning layer serves as an ignition source for the next adjacent layer, and so on. Combustion terminates when equilibrium is achieved between the total heat energies of the reactants and the total heat energies of the products. Most reactions terminate when what is called thermal equilibrium has been attained—i.e. when the energy of the reactants equals the energy of the products.

A relationship exists between the ignition temperature and the pressure of the mixture under specific conditions. The figure shows the relationship for a mixture of hydrogen and oxygen. Only one temperature corresponds to a given pressure, whereas one or three pressures, called the explosion limits, may correspond to one temperature. The mechanism of the reaction determines the explosion limits: the reaction can proceed only when the steps in the sequence of reactions occur faster than the terminal steps. Thus, for combustion to be initiated with light or with a spark, the light intensity or the spark energy must exceed certain minimal values.

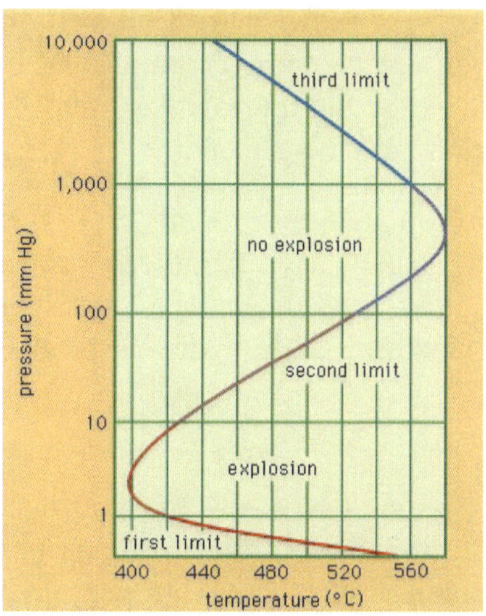

Explosion limits of a hydrogen-oxygen mixture.

The complexity of the combustion reaction mechanism and the rapidly varying temperatures and concentrations in the mixture make it difficult and often impossible to derive an equation that would be useful for predicting combustion phenomena over wide temperature and concentration ranges. Instead, use is made of empirical expressions derived for specific reaction conditions.

Special Combustion Reactions

Reactions of oxygen with hydrogen, with carbon monoxide, and with hydrocarbons are most important from the standpoint of theory and, at the same time, are of great practical value.

Hydrogen combustion proceeds by complicated branched-chain reactions involving the interaction of hydrogen and oxygen atoms with oxygen and hydrogen molecules, respectively, to produce hydroxyl radicals. The final reaction product is water, formed by the combination of hydroxyl with hydrogen molecules. Reactions that terminate the chain, such as recombination of atoms and fragments of molecules, and the adsorption of active particles on solid surfaces, also play an important part in the mechanism of hydrogen combustion. The knowledge of rate constants for these processes derived empirically makes possible a quantitative description of all combustion characteristics, such as explosion limits, delay of ignition, and burning velocity.

Combustion of carbon monoxide is mainly restricted to mixtures containing hydrogen or hydrogen compounds. In this case the reaction mechanism differs from that of hydrogen combustion chiefly in that it involves a step of fast interaction between hydroxyl and carbon monoxide.

Pure mixtures of carbon monoxide and oxygen (or air) can be ignited only with sparks of high energy or under high pressures and temperatures. The chemical mechanism of their combustion is not yet clear, probably because oxidation of carbon monoxide, a reaction that is part of the combustion of practically all natural fuels, usually occurs in the presence of hydrogen or hydrogen compounds: the breakdown of wood, coal, petroleum, etc., during burning produces carbon monoxide, hydrogen, and compounds of carbon and hydrogen.

The mechanisms of combustion of hydrocarbons and of other organic compounds are known in general outline only. Many elementary steps of hydrocarbon combustion involving oxygen and hydrogen atoms, hydroxyl, and organic radicals are similar to those for hydrogen; however, the overall mechanism of hydrocarbon combustion is complicated by the diversity of molecules and radicals involved. Moreover, with oxidation, thermal decomposition, a breakdown of complex organic compounds without oxidation, also takes place.

Two types of hydrocarbon combustion have been defined: (1) slow combustion at temperatures below 500 °C, including cool flames observed at certain pressures, and (2) combustion at higher temperatures, accompanied by hot flames. Ignition in two stages is characteristic of higher hydrocarbons, in which a cool-flame stage yielding readily oxidizable products precedes that of the hot flame.

Physical Processes

In addition to chemical reactions, physical processes that transfer mass and energy by diffusion or convection occur in gaseous combustion. In the absence of external forces, the rate of component diffusion depends upon the concentration of the constituents, pressure, and temperature changes, and on diffusion coefficients (a measure of the speed of diffusion). The latter are either measured or calculated in terms of the kinetic theory of gases. The process of diffusion is of great importance in combustion reactions, in flames—that is, in gaseous mixtures, and in solids or liquids. Diffusion heat transfer (by molecular means) follows a law (Fourier's law) stating that the heat flux (a measure of the quantity of heat) is proportional to the temperature gradient (the difference in temperature between two limiting temperatures). The coefficient of proportionality, called the thermal conductivity coefficient, is also measured or calculated in terms of the kinetic theory of gases, like the diffusion coefficient.

Convective transport of mass and energy may be accounted for by buoyant forces, external forces, and turbulent and eddy motions. Convection is in the main responsible for the mixing of gases (e.g., in furnaces).

Sublimation (the direct evaporation of a solid without intermediate liquid stage), evaporation of liquids, and the mechanical destruction of burning samples, all methods of transforming solids and liquids into gases, greatly contribute to their ease of combustion. In general, the combustion of condensed systems—i.e. liquids and solids—is more complex than that of gases, which accounts for the greater development of the gas combustion theory.

Energy transport in combustion may also occur by the emission of light, mostly in the infrared.

Combustion Phenomena and Classification

All flames can be classified either as premixed flames or as flames that burn without premixing.

Premixed Flames

Flame combustion is most prominent with fuels that have been premixed with an oxidant, either oxygen or a compound that provides oxygen, for the reaction. The temperature of flames with this mixture is often several thousand degrees. The chemical reaction in such flames occurs within a narrow zone several micrometers thick. This combustion zone is usually called the flame front.

Dilution of the burning mixture with an inert gas, such as helium or nitrogen, lowers the temperature and, consequently, the reaction rate. Great amounts of inert gas extinguish the flame, and the same result is achieved when substances that remove any of the active species are added to the flame. Conditions must be such that the flame is fixed at the burner nozzle or in the combustion chamber; this positioning is required in many practical uses of combustion. Various devices, such as pilot flames and recirculation methods, are designed for this purpose.

The principal quantitative characteristic of a flame is its normal, or fundamental, burning velocity, which depends on the chemical and thermodynamic properties of the mixture and on pressure and temperature, under given conditions of heat loss. The burning velocity value ranges from several centimeters to even tens of meters per second. The dependence of the burning velocity on molecular structure, which is responsible for fuel reactivity, is known for a great many fuel-air mixtures.

A widely applied thermal theory, one of the first flame propagation theories, implies that combustion proceeds primarily at temperatures close to the maximum the flame can achieve. A set of differential equations developed for thermal conductivity and diffusion is reduced to one equation that yields the burning velocity value. Further development of the theory has been made, and computers now make most simplifications unnecessary.

According to thermal theories, flame propagation is accounted for by heat energy transport from the combustion zone to the unburned mixture, which raises the temperature of the mixture. Diffusion theory assumes that thermodynamic equilibrium sets in behind the flame front at a maximum temperature and that radicals produced in this zone diffuse into the unburned mixture and ignite it. Both heat transport and diffusion of active particles must be considered essential for ignition.

Diffusion Flames

Diffusion flames, smoothly flowing (laminar) or turbulent, belong to the class of flames whose ingredients are not mixed prior to entering the burning zone. Molecular or turbulent diffusion is responsible for the mixing of the gases in such flames. The distribution of the combustible material and of oxygen over various flame cross sections is regular in laminar flames but becomes much more complicated in turbulent gas flows. The transition from laminar to turbulent flames occurs at a certain point of the flow and depends upon the actual conditions.

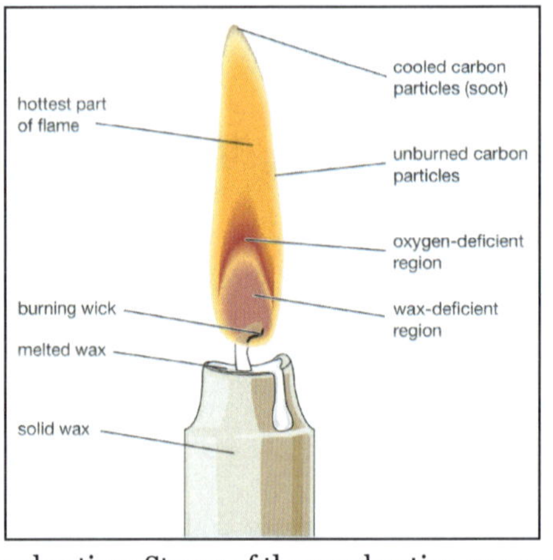

Combustion: Stages of the combustion process.

Industrial flames, including those activated in furnaces, belong to the turbulent diffusion type. The theory of diffusion flames, however, is less advanced than that of premixed flames and, since gas intermixing is mainly responsible for the structure and the features of diffusion flames, the theory operates more in terms of physics and thermodynamics than chemical reactions. Flames such as those of candles, of liquid droplets, and of many propellants and condensed fuels are diffusion flames.

Oxidizing and Reducing Flame

When a premixed flame burns in open air with an excess of fuel, there appears in addition to the flame zone a zone of diffusion flame; this is accounted for by the diffusion of atmospheric oxygen, as, for example, in the Bunsen flame produced by a burner to which the air intake can be regulated, thereby altering the flow from an intensely hot one—in which most of the fuel gases are oxidized to carbon dioxide and water—to a relatively low temperature flux in which most of the fuel gas is only partially oxidized. Such flames consist of an inner core and an outer cone—two zones in which different chemical reactions take place, the reducing and oxidizing zones, respectively. The oxidizing nature of the outer cone is due to excess oxygen.

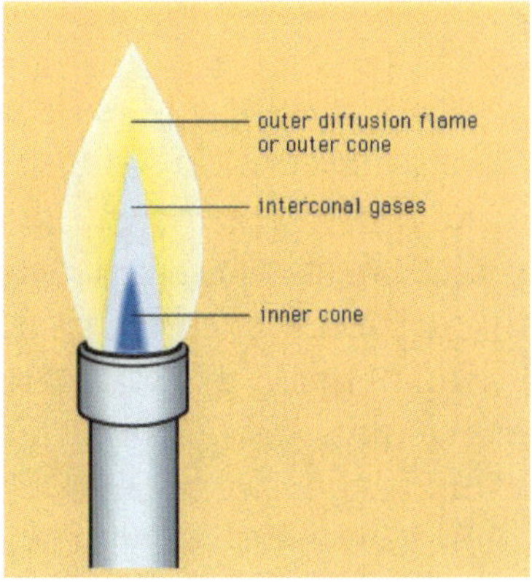

Cone of a Bunsen burner flame.

Explosions

The transition from combustion to explosion is caused by an acceleration of the reaction, induced either by a rise in temperature or by increasing lengths of the reaction chain. The first is called thermal explosion, and the second is called chain explosion.

Thermal Explosions

Thermal explosion theory is based on the idea that progressive heating raises the rate at which heat is released by the reaction until it exceeds the rate of heat loss from the area. At a given composition of the mixture and a given pressure, explosion will occur at a specific ignition temperature that can be determined from the calculations of heat loss and heat gain.

The thermal explosion theory accounts for temperature rise and fuel consumption during the induction period. At sufficiently high rates of consumption the explosion will not occur.

Chain-branch Reactions

It follows from the theory of branched-chain reactions that there is a limit to ignition, or to explosion, without a rise of temperature. In this case, what is called a chain explosion will occur when the probabilities of chain branching and of termination are equal. Usually, however, explosions are of a chain-thermal nature (i.e., both heat accumulation and chain auto-acceleration contribute to explosion).

Detonation

The progressive acceleration of reaction accounted for by the flame front area advancing at a supersonic velocity and by transition from laminar to turbulent flow gives rise to a shock wave. The increase in temperature due to compression in the shock wave results in self-ignition of the mixture, and detonation sets in. The shock wave–combustion zone complex forms the detonation wave. Detonation differs from normal combustion in its ignition mechanism and in the supersonic velocity of 2–5 kilometers per second for gases and 8–9 kilometers per second for solid and liquid explosives.

Detonation is impossible when the energy loss from the reaction zone exceeds a certain limit. The detonation limits observed for high explosives are also eventually dependent on factors responsible for the chemical reaction rate: For example, the charge and diameter of the grain.

Special Aspects

Emission of light is a characteristic feature of combustion. Infrared, visible, and ultraviolet bands of molecules and atomic lines are usually observed in flame spectra. Moreover, continuous spectra from incandescent particles or from recombination of atoms, radicals, and ions are frequently observed. The sources of flame radiation are the thermal energy of gas (thermoluminescence) and the chemical energy released in exothermic elementary reactions (chemiluminescence). In a Bunsen burner fed with a sufficient amount of air, up to 20 percent of the reaction heat is released as infrared energy and less than 1 percent as visible and ultraviolet radiation, the infrared being mostly thermoluminescence. Radiation from the inner cone of the Bunsen flame in the visible and ultraviolet regions represents chemiluminescence.

Many flames contain electrons and positive and negative ions, a fact evident from the electrical conductivity of flames, and also from the deviation of the flame cone in an electric field (the charges are attracted or repelled, distorting the flame), a phenomenon usually interpreted as a mechanical effect called electric wind. The resulting change of the flame shape can affect the burning velocity. Ionization, like the emission of light, can be the result of equilibrium processes, when it is called thermal ionization, or it can be related to chemical processes and called chemical ionization. Thermal ionization may be expected in very hot flames containing alkali metals or alkaline-earth metals (for example, sodium and calcium) as impurities because of their low-ionization potentials. The high concentrations of ions and electrons in the flames of organic species are undoubtedly due to chemical, rather than to thermal, ionization. This is seen in the fact that electroconductivity in the inner cone of the Bunsen flame is several times higher than that of the outer cone. The reactions of ions and electrons may yield atoms and radicals and in this way affect the burning velocity.

Formation of soot is a feature of all hydrocarbon flames. It makes the flame luminous and nontransparent. The mechanism of soot formation is accounted for by simultaneous polymerization, a process whereby molecules or molecular fragments are combined into extremely large groupings, and dehydrogenation, a process that eliminates hydrogen from molecules.

Applications

The uses of combustion and flame phenomena can be categorized under five general heads.

Heating Devices

Heating devices for vapour production (steam, etc.), in metallurgy, and in industry generally, utilize the combustion of gases, wood, coal, and liquid fuels. Control of the combustion process to obtain optimal efficiency is ensured by proper ratio and distribution of the fuel and the oxidant in the furnace, stove, kiln, etc., by choice of conditions for heat transport from the combustion products to the heated bodies, and by appropriate aerodynamics of gas flows in the furnace. Radiation contributes to a certain extent to heat exchange. Combustion in furnaces being a very complicated process, only general ideas can be given by the combustion theory, so that the optimal conditions and the furnace design are usually decided empirically.

Explosives

The combustion and detonation of explosives are widely used in all sorts of work with mechanical action or explosion as the eventual goals. Practical applications of explosives are based on the theory of their combustion and detonation. The combustion of condensed explosives occurs mostly in the gas phase because of their evaporation, sublimation, or decomposition and can be treated in terms of the theory of gas combustion, which provides for the burning velocity, its dependence on temperature and pressure, and the parameters determining the combustion regime and the nature of explosives. Control of combustion and detonation in their practical applications is made possible by use of the theory, together with the results of experimental investigations on combustion and detonation.

Internal-combustion Engines

These comprise various engines, gas turbines, turbojets, and ramjets. The Otto engine operates with a mixture compressed in a cylinder by a piston. Shortly before the piston reaches the top the mixture is ignited with a spark, and the flame propagates at a normal velocity into the unburned mixture, increasing the pressure and moving the piston. There is a maximum of compression for any mixture composition and any engine design. Detonation occurs beyond this maximum because of the appearance of centres where self-ignition takes place before the flame front. Loss of power is one result of detonation; compounds hindering self-ignition are used to prevent it.

The diesel engine operates with a fuel spray injected into the engine cylinder as liquid droplets that mix with air by turbulent diffusion and evaporate. At normal operations of the engine the temperature of compressed air is sufficiently high for self-ignition of the fuel.

In gas turbines, compressed air enters the combustion chamber where it mixes with the fuel. The expanding combustion products impart their energy to the turbine blades.

Two kinds of jet engines are used in aircraft: the turbojet and the ramjet. The turbine of a turbojet engine is used to operate the compressor. Thrust comes from the repulsion of products flowing out of a nozzle. In a ramjet engine, air is compressed and slowed down in the diffuser without any compressor or turbine device.

Rocket Propulsion

The products of combustion of gaseous, liquid, or solid propellants in rockets are ejected from the combustion chamber through the (de Laval) nozzle at a high velocity. Knowledge of the kinetics of chemical processes in the nozzle is essential to determine the thrust required. The thrust decreases with the increasing mean molecular weight of the combustion products. Mixtures of low molecular weight and high heat of combustion, therefore, are used for rockets.

Chemical Reactions

Flames are used in various ways to produce chemical reactions. The bead test in analytical chemistry is one example. The reducing power of a flame that has insufficient oxygen is utilized in limited ways. The soot produced by some flames is commercially useful, and the manufacture of coke and charcoal is dependent on the judicious control of combustion and flame.

References

- Combination-Reactions, Chemical-Reactions, Introductory-Chemistry: libretexts.org, Retrieved 2 May, 2019
- Combination-Reactions, Chemical-Reactions-and-Interactions, General-Chemistry: libretexts.org, Retrieved 16 February, 2019
- Combination-reaction-examples: zohosites.com, Retrieved 7 January, 2019
- Definition-of-decomposition-reaction: thoughtco.com, Retrieved 21 July, 2019
- Decomposition-Reactions, Chemical-Reactions, Introductory-Chemistry: libretexts.org, Retrieved 19 March, 2019
- Replacement-Reactions-MS-PS, single-replacement-reactions, chemistry: ck12.org, Retrieved 23 March, 2019
- Single-replacement-reactions, types-of-chemical-reactions, chemical-reactions-stoichiome, chemistry, science: khanacademy.org, Retrieved 1 February, 2019
- Single-Replacement-Reactions, Chemical-Reactions, Introductory-Chemistry: libretexts.org, Retrieved 30 April, 2019
- Double-replacement-reactions, types-of-chemical-reactions, chemical-reactions-stoichiome, chemistry, science: khanacademy.org, Retrieved 12 January, 2019
- Acid-base-reaction: newworldencyclopedia.org, Retrieved 22 April, 2019
- Oxidation-reduction-reaction, science: britannica.com, Retrieved 28 August, 2019
- Combustion, science: britannica.com, Retrieved 3 June, 2019

6

Chemistry of Group Elements

Elements within the periodic table are arranged on the basis of their properties into different groups. A few of these groups are alkali metals, alkaline earth metals, boron group and carbon family. The topics elaborated in this chapter will help in gaining a better perspective about the different groups of elements.

The Group 1 Elements

Alkali Metals

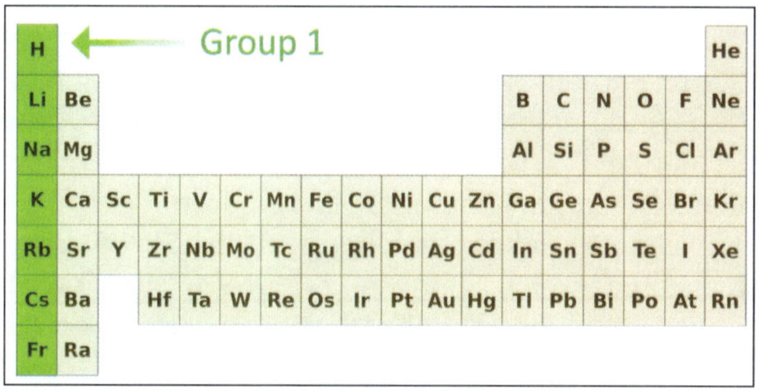

Included in the Group 1 of the periodic table are the following elements:

- Lithium
- Sodium
- Potassium
- Rubidium
- Caesium

The general electronic configuration of Group 1 elements is ns^1. They have a strong tendency to donate their valence electron in the last shell to form strong ionic bonds. They have the least nuclear charge in their respective periods. As we move down the group, the atomic radius increases. Therefore, the nuclear charge decreases. Caesium is the most metallic element in the group.

In order to prevent the elements from coming in contact with oxygen, they are stored in jars that contain oil. The melting points of these elements are quite low, which is 180° Celsius in the case of Lithium, while it is 39° Celsius in the case of Rubidium. When it comes to the density of the metal, group one elements display a very low level of density of up to 1 gcm-3 which means that they can easily float on the surface of the water.

Therefore, if we decide to cut these metals, we will be able to do so, without much trouble. Upon being cut into two halves, we can observe that their surface is as shiny as any other metal but even after they are stored in oil, they undergo tarnishing. As a conductor of heat and electricity, they are excellent.

General Properties of the Group I

Physical Properties

The alkali metals have the high thermal and electrical conductivity, lustre, ductility, and malleability that are characteristic of metals. Each alkali metal atom has a single electron in its outermost shell. This valence electron is much more weakly bound than those in inner shells. As a result, the alkali metals tend to form singly charged positive ions (cations) when they react with nonmetals. The compounds that result have high melting points and are hard crystals that are held together by ionic bonds (resulting from mutually attractive forces that exist between positive and negative electrical charges). In the metallic state, either pure or in alloys with other alkali metals, the valence electrons become delocalized and mobile as they interact to form a half-filled valence band. As with other metals, such a partially filled valence band is a conduction band and is responsible for the valence properties typical of metals. In passing from lithium to francium, the single electron tends to be less strongly held. Generally, the energy necessary to remove the outermost electron from the atoms of an element, the ionization energy, decreases in the periodic table toward the left and downward in each vertical file, with the result that the most easily ionizable element in the entire table is francium, followed closely by cesium. The alkali metals, which make up the extreme left-hand file, have ionization energies ranging from 124.3 kilocalories per mole (kcal/mole) in lithium to 89.7 kcal/mole in cesium (omitting the rare radioactive element francium). The alkaline-earth metals, the next group to the right, have higher ionization energies ranging from 214.9 in beryllium to 120.1 kcal/mole in barium.

The electronegativity scale of the elements compares the ability of the atoms of the various elements to attract electrons to themselves. In the periodic table the electronegativities range from 0.7 for cesium, the least electronegative of the elements, to 4.0 for fluorine, the most electronegative. Metals are ordinarily considered to be those elements having values less than 2.0 on the electronegativity scale. As a group the alkali metals are the least electronegative of the elements, ranging from 0.7 to 1.0 on the scale, while the alkaline earths, the next group on the table, have electronegativities ranging from about 0.9 to 1.5.

The table summarizes the important physical and thermodynamic properties of the alkali metals. At atmospheric pressure these metals are all characterized by a body-centred cubic crystallographic arrangement (a standard pattern of atoms in their crystals), with eight nearest neighbours to each atom. The closest distance between atoms, a characteristic property of crystals, increases with increasing atomic weight of the alkali metal atoms. As a group, the alkali metals have a looser crystallographic arrangement than any of the other metallic crystals, and cesium—because of its greater atomic weight—has an interatomic distance that is greater than that of any other metal.

Properties of the Alkali Metals

	Lithium	Sodium	Potassium	Rubidium	Cesium	Francium
Atomic number	3	11	19	37	55	87
Atomic weight (or stablest isotope)	6.941	22.99	38.098	86.468	132.905	223
Color of element	silver	silver	silver	silver	silver	—
Melting point (°C)	180.5	97.72	63.38	39.31	28.44	27
Boiling point (°C)	1,342	883	759	688	671	677
Density at 20 °C (grams per cubic centimetre)	0.534	0.971	0.862	1.532	1.873	—
Volume increase on melting (percent)	1.51	2.63	2.81	2.54	2.66	—
Valence	1	1	1	1	1	1
Mass number of most common isotopes (terrestrial abundance, percent)	6 (7.59), 7 (92.41)	23 (100)	39 (93.2581), 40 (0.0117), 41 (6.7302)	85 (72.17), 87 (27.83)	133 (100)	—
Color imparted to flame	red	yellow	violet	yellow violet	blue	—
Main spectral emission lines (wavelength, angstroms)	6,708; 6,104	5,890; 5,896	7,699; 7,665	4,216; 4,202	4,593; 4,555	—
Heat of fusion (calories per mole/ kilojoules per mole)	720 (3)	621 (2.6)	557 (2.33)	523 (2.19)	500 (2.09)	500 (2)
Specific heat (joules per gram kelvin)	3.582	1.228	0.757	0.363	0.242	—
Electrical resistivity at 293–298 K (microhm-centimetres)	9.5	4.9	7.5	13.3	21	—
Magnetic susceptibility(cgs units)	14.2 (10–6)	16 (10–6)	20.8 (10–6)	17 (10–6)	29 (10–6)	—
Crystal structure	body-centred cubic	body-centred cubic	body-centred cubic	body-centred cubic	body-centred cubic	—
Radius: atomic (angstroms)	1.67	1.9	2.43	2.65	2.98	—
Radius: ionic (+1 ion, angstroms)	0.9	1.16	1.52	1.66	1.81	1.94
Radius: metallic (angstroms, 12-coordinate)	1.57	1.91	2.35	2.5	2.72	2.8
First ionization energy(kilojoules per mole)	520.2	495.8	418.8	403	375.7	380
Oxidation potential for oxidation from the 0 to +1 oxidation state at 25 °C (volts)	3.04	2.71	2.93	2.92	2.92	2.92
Electronegativity (Pauling)	0.98	0.93	0.82	0.82	0.79	0.7

Vapour-pressure data for the alkali metals and for two alloys formed between elements of the group show that the vapour pressures increase in regular fashion with increasing atomic weight. Cesium is the most volatile of the alkali metals, with a boiling point of 671 °C (1,240 °F). The boiling points of the alkali metals decrease in regular fashion as the atomic numbers increase, with the highest, 1,317 °C (2,403 °F), being that of lithium.

The melting points of the alkali metals as a group are lower than those of any other nongaseous group of the periodic table, ranging between 179 °C (354 °F) for lithium and 28.5 °C (83.3 °F) for cesium. Among the metallic elements, only mercury has a lower melting point (−38.9 °C, or −38.02 °F) than cesium. The low melting points of the alkali metals are a direct result of the large interatomic distances in their crystals and the weak bond energies associated with such loose arrays. These same factors are responsible for the low densities, low heats of fusion, and small changes in volume upon fusion of the metals. Lithium, sodium, and potassium are less dense than water.

The large size of an alkali metal atom (and the resulting low density of the metal) results from the presence of only one, weakly bound electron in the large outer s-type orbital. Upon going from the noble-gas configuration of argon (atomic number 18) to potassium (atomic number 19), the added electron goes into the large 4s orbital rather than the smaller 3p orbital. When, however, potassium, rubidium, or cesium metals are subjected to increasing pressure (up to one-half million atmospheres or more), a number of phase transitions occur. Ultimately, occupation of a d-type orbital becomes preferred over that of the s-orbital, with the result that these alkali metals resemble transition metals. Under such circumstances, alloys with transition metals (such as iron) can form, a result that does not occur at low pressures. It has been proposed that the lower-than-expected density of Earth's core may be the result of the formation of a potassium-iron alloy under the extreme pressures that occur there.

The alkali metals have played an important role in quantum physics. Some alkali metal isotopes, such as rubidium-87, are bosons. Dilute atomic gases of such alkali metal isotopes, confined by magnetic fields or "laser mirrors" and cooled to temperatures near absolute zero, form Bose-Einstein condensates. In this state, the cluster of atoms is in a single quantum state and exhibits macroscopic behaviors normally seen only with atomic-sized particles. These include interference effects and coherent motion of the entire "cloud" of atoms.

Chemical Properties

Since the alkali metals are the most electropositive (the least electronegative) of elements, they react with a great variety of nonmetals. In its chemical reactivity, lithium more closely resembles Group 2 (IIa) of the periodic table than it does the other metals of its own group. It is less reactive than the other alkali metals with water, oxygen, and halogens and more reactive with nitrogen, carbon, and hydrogen.

Reactions with Oxygen

The alkali metals tend to form ionic solids in which the alkali metal has an oxidation number of +1. Therefore, neutral compounds with oxygen can be readily classified according to the nature of the oxygen species involved. Ionic oxygen species include the oxide, O^{2-}, peroxide, O_2^{2-}, superoxide, O_2^-, and ozonide O_3^-. Compounds that can be prepared that contain an alkali metal, M, and oxygen are therefore the monoxide, M_2O, peroxide, M_2O_2, superoxide, MO_2, and ozonide, MO_3. Rubidium and cesium and, possibly, potassium also form the sesquioxide, M_4O_6, which contains two peroxide anions and one superoxide anion per formula unit. Lithium forms only the monoxide and the peroxide.

All the alkali metals react directly with oxygen; lithium and sodium form monoxides, Li_2O and Na_2O, and the heavier alkali metals form superoxides, MO_2. The rate of reaction with oxygen, or with air, depends upon whether the metals are in the solid or liquid state, as well as upon the degree of mixing of the metals with the oxygen or air. In the liquid state, alkali metals can be ignited in air with ease, generating copious quantities of heat and a dense choking smoke of the oxide.

The free energy of formation (a measure of stability) of the alkali metal oxides at 25 °C (77 °F) varies widely from a high of −133 kcal/mole for lithium oxide to −63 kcal/mole for cesium oxide. The close approach of the small lithium ion to the oxygen atom results in the unusually high free energy of formation of the oxide. The peroxides (Li_2O_2 and Na_2O_2) can be made by passing oxygen through a liquid-ammonia solution of the alkali metal, although sodium peroxide is made commercially by oxidation of sodium monoxide with oxygen. Sodium superoxide (NaO_2) can be prepared with high oxygen pressures, whereas the superoxides of rubidium, potassium, and cesium can be prepared directly by combustion in air. By contrast, no superoxides have been isolated in pure form in the case of lithium or the alkaline-earth metals, although the heavier members of that group can be oxidized to the peroxide state. The cyanides of potassium, rubidium, and cesium, which are less stable than the lower oxides, can be prepared by the reaction of the superoxides with ozone.

Reactions with Water

The alkali metals all react violently with water according to $M + H_2O \rightarrow MOH + 1/2\ H_2$. The rate of the reaction depends on the degree of metal surface presented to the liquid. With small metal droplets or thin films of alkali metal, the reaction can be explosive. The rate of the reaction of water with the alkali metals increases with increasing atomic weight of the metal. With the heavier alkali metals, the hydroxides are highly soluble; thus, they are removed readily from the reacting surface, and the reaction can proceed with unabated vigour. The reaction involves equimolar mixtures (that is, equal numbers of atoms or molecules) of the alkali metal and water to form a mole (an amount equal to that of the reactants) of alkali metal hydroxide and half a mole of hydrogen gas. These reactions are highly exothermic (give off heat), and the hydrogen that is generated can react with oxygen to increase further the heat that is generated.

Reactions with Nonmetals

Of the alkali metals, only lithium reacts with nitrogen, and it forms a nitride(Li_3N). In this respect it is more similar to the alkaline-earth metals than to the Group 1 metals. Lithium also forms a relatively stable hydride, whereas the other alkali metals form hydrides that are more reactive. Lithium forms a carbide (Li_2C_2) similar to that of calcium. The other alkali metals do not form stable carbides, although they do react with the graphite form of carbon to give intercalation compounds (substances in which the metal atoms are inserted between layers of carbon atoms in the graphite structure).

The alkali metals can be burned in atmospheres of the various halogens to form the corresponding halides. The reactions are highly exothermic, producing up to 235 kcal/mole for lithium fluoride. The alkali metals react with nonmetals in Groups 15 and 16 of the periodic table. Sulfides can be formed by the direct reaction of the alkali metals with elemental sulfur, furnishing a variety of sulfides. Phosphorus combines with the alkali metals to form phosphides with the general formula M_3P.

Formation of Alloys

The characteristics of alloy behavior in alkali metals can be evaluated in terms of the similarity of the elements participating in the alloy. Elements with similar atomic volumes form solid solutions (that is, mix completely in all proportions); some dissimilarity in atomic volumes results in eutectic-type systems (solutions formed over limited concentration ranges), and further dissimilarity results in totally immiscible systems. The high-pressure transition in potassium, rubidium, and cesium that converts these *s*-type metals to more transition metal-like *d*-type metals yields atomic volumes that are similar to those of many transition metals at the same pressure. This permits alloys or compounds to form between these alkali metals and such transition metals as nickel or iron.

The elements potassium, rubidium, and cesium, which have rather similar atomic volumes and ionization energies, form complete solid solutions and mixed crystals. Sodium, which is a significantly smaller atom than potassium and has a higher ionization energy, tends to form eutecticsystems with potassium, rubidium, and cesium. Even greater dissimilarity exists in the atomic volumes of sodium and lithium, resulting in insolubilities of the liquid phases. The consolute temperature (the temperature at which the two liquids become completely miscible) increases on going from the lithium-sodium alloy system to the lithium-cesium system. Lithium and cesium can coexist as two separate liquid phases at temperatures up to at least 1,100 °C (2,000 °F).

There is only one example of solid miscibility in alkali–alkaline-earth-metal binaries—the lithium-magnesium system, in which the two elements are very similar. Sodium forms compounds only with barium in the alkaline-earth-metal series. The heavier alkali metals all tend to form immiscible liquid phases with the alkaline earth metals.

Several elements in Group 12 (IIb) of the periodic table (zinc, cadmium, and mercury) react with the alkali metals to form compounds. Mercury forms at least six compounds, commonly termed amalgams, with each of the five alkali metals, and with the exception of the amalgam with lithium, the highest melting point compound in each series has the formula MHg_2. Lithium and sodium also form compounds with cadmium and zinc.

Formation of Complexes

Until the late 1960s there were few complexes of the alkali metal cations with organic molecules. Specialized biological molecules such as valinomycin were known to complex selectively the potassium cation K^+ for transport across cell membranes, but synthetic ionospheres (molecules that can form complexes with ions) were rare. All the alkali cations have a charge of +1 and, except for lithium, are chemically similar and rather inert. The only significant difference between one alkali cation and another is the size.

The synthesis of crown ethers by American chemist Charles J. Pedersen in 1967 provided size-selective cyclic molecules consisting of ether oxygens forming a ring or "crown" that could complex a cation of the right size to fit into the hole in the centre of the molecule. In some cases two ether molecules can encapsulate a cation in a "sandwich" fashion. For example, K^+ just fits into the centre of an 18-crown-6 ring (18 atoms in the ring, 12 of which are carbon atoms and 6 are ether oxygen atoms) to form a 1:1 complex (that is, 1 cation:1 crown ether), $K^+(18C6)$. Cs^+ is too large to fit into the ring but can be complexes on one side to form the $Cs^+(18C6)$ complex or can be sandwiched between two 18-crown-6 molecules to form the 1:2 complex, $Cs^+(18C6)_2$. Thus, the selectivity of a crown ether for a particular cation depends on the ring size. Common crown ethers are 12-crown-4, 15-crown-5, and 18-crown-6. These molecules are selective for Li^+, Na^+, and K^+, respectively.

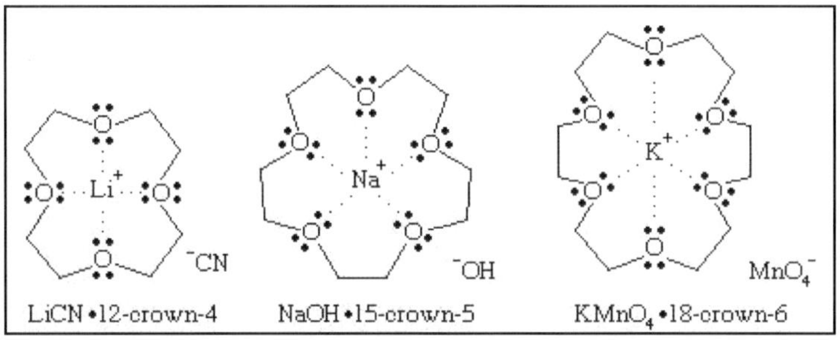

Three crown ethers, each with a complexed alkali metal cation.

Even greater affinity for alkali cations was achieved by the synthesis of cryptands by French chemist Jean-Marie Lehn in 1968 and spherands by American chemist Donald Cram in 1979. These are three-dimensional molecules with an internal cavity or crypt that can completely encapsulate the alkali cation. By synthesizing molecules with different cavity sizes, the selectivity for particular cations over those of the "wrong" size to fit in the cavity can be controlled. It should be noted, however, that these molecules are not rigid and that flexibility of the framework can alter the cavity size to accommodate alkali cations of different sizes, although with differences in the strength of complexation.

Since the initial syntheses of crown ethers and cryptands, thousands of complexants for cations of various sizes, charges, and geometries have been synthesized. This has led to an entirely new branch of chemistry called supramolecular chemistry.

Analytical Chemistry of the Alkali Metals

Classical methods of separation and analysis of alkali metals are rather difficult and time consuming. For lithium they include such procedures as selective extraction of lithium chloride into organic solvents and the detection of lithium with azo dyes that give highly sensitive Color reactions in alkaline solutions. A modification of the uranyl acetate test (the precipitation of an insoluble sodium salt with uranyl acetate) has been used as a standard test for the presence of sodium. The use of a cobaltinitrite solution permits separation of potassium from sodium by precipitation of the insoluble potassium salt. There are essentially no satisfactory analytical methods for rubidium and cesium based on the use of reagents in solution.

Classical methods of separation of the alkali metals have been largely supplanted by chromatographic elution. Strongly acidic cation-exchange resins and aqueous acidic solutions are used. Generally the affinity increases with atomic weight so that the ions are eluted in the order $Fr^+ > Cs^+ > Rb^+ > K^+ > Na^+ > Li^+$, which is the order of decreasing size of the hydrated ions. Ion-exchange resins that are specific for lithium have been developed. Macrocyclic compounds such as crown ethers and cryptands that are selective for particular alkali metal ions have been synthesized. They form cationic complexes that can be dissolved in organic solvents such as chloroform ($CHCl_3$) with counterions such as picrate ($C_6H_2[NO_2]_3O^-$).

The characteristic flame Colors of the alkali metals (red, yellow, violet, red, and blue for Li, Na, K, Rb, and Cs, respectively) are qualitative indicators of the modern analytical methods used to determine the concentrations of alkali-metal salts in aqueous solution. The intensities of the characteristic spectral lines in emission after excitation by a flame or ICP (inductively coupled plasma) give quantitative measures of the individual alkali metal concentration in the parts per million range or lower. Determination of one alkali metal in the presence of another, however, can result in interference, which can be reduced by using specially prepared standard solutions that contain known amounts of the interfering metals.

The analysis of the alkali-metal samples for the presence of nonmetallic elements, such as oxygen, carbon, hydrogen, and nitrogen, requires specialized techniques. The oxygen content of sodium and potassium samples can be determined by extraction of the free alkali metal with mercury, leaving behind mercury-insoluble oxides and carbonates, which can subsequently be analyzed by means of solution methods. The oxygen content of rubidium and cesium can be accurately determined by precise measurement of the freezing point of these two elements.

The carbon content of alkali metals can be analyzed by oxidation of the alkali metal in pure oxygen, followed by infrared measurement of the carbon dioxide generated during combustion. For the analysis of nitride in lithium, the nitride commonly is converted to ammonia, and the ammonia is measured by colorimetric analysis.

The Group 2 Elements

The Group 2 alkaline earth metals include Beryllium, Magnesium, Calcium, Barium, Strontium and Radium and are soft, silver metals that are less metallic in character than the Group 1 Alkali Metals. Although many characteristics are common throughout the group, the heavier metals such as Ca, Sr, Ba, and Ra are almost as reactive as the Group 1 Alkali Metals. All the elements in Group 2 have two electrons in their valence shells, giving them an oxidation state of +2.

Physical Properties of Alkaline Earth Metals

1. Physical nature:

   These elements have two electrons in their outermost orbital. They have a silvery luster. But, it soon disappears upon exposure to air. They are malleable and ductile but very less when compared to alkali metals.

2. Atomic Volume and Radius:

   As we go down the group, the number of electrons is less when compared to the increase in the number of available orbitals. Hence, the atomic radius increases gradually as we move down the group. It should be noted that these elements are smaller when compared to alkali metals. This is because of the presence of two electrons in the outermost shell. Hence, the effect of the nucleus on the outermost shell is comparatively more than the alkali metals which have a single electron.

3. Density:

   Due to presence of two electrons in the outermost shell they can be more densely packed compared to alkali metals.

4. Melting and Boiling Points:

   These elements have a higher boiling and melting points. Due to presence of two electrons in the valence shell they remain tightly packed in solid state.

5. Ionization Energy:

   Alkaline earth metals have a smaller size and higher nuclear charge as a result of which the valence electrons are held strongly. Hence, more amount of energy is required to remove an electron from the valence electron resulting in high ionization energies.

6. Oxidation State:

   Due to the presence of two electrons in the valence shell, elements of group IIA exhibit +2 oxidation i.e. they are bivalent.

7. Electropositivity:

   As the size to charge ratio is very high, they are highly electropositive in nature.

8. Electronegativity:

   As they are highly electropositive, they are less electronegative. Electronegativity decreases as we move down the group. Beryllium is highly electronegative due to small size.

9. Conductivity:

   They are good conductors of heat and electricity. This is due to the presence of a two electrons that can very easily move within the crystal lattice of the elements.

10. Flame colorization:

    When the elements are heated, electrons present in the valence shell are excited to higher energy levels. When the excited electrons return back after losing the energy they emit certain amount light. All Alkaline earth metals except beryllium and magnesium emit various colors depending on the degree of excitation. These two elements due to their smaller size have high ionization energies and high excitation energy. Hence, they are not excited to higher energy states and thus no flame colorization.

11. Reducing property:

    The two electrons in the valence shell can be very easily removed. Hence, all alkaline earth metals act as strong reducing agents. The reducing nature increases as we move down the group. However, the reducing nature of alkaline earth metals is less than their fellow s-block elements, the alkali metals.

Chemical Properties of Alkaline Earth Metals

1. Reactivity towards air and water- Beryllium and magnesium do not react with oxygen and water. They are inert towards water and oxygen as they form oxide layer on their surface. But powdered beryllium burns in air to give BeO and Be_3N_2. Magnesium burns in air to give

MgO and Mg_3N_2. Calcium, strontium, and barium are easily oxidized in air to form oxides and nitrides. They also react with water to form hydroxides.

$$2Mg_{(s)} + O_{2(g)} \rightarrow 2MgO_{(s)}$$
$$2Ca_{(s)} + O_{2(g)} \rightarrow 2CaO_{(s)}$$

2. Reactivity towards halogens- Like alkali metals, all alkaline earth metals reacts with halogen to form halides.

$$Be(s) + Cl_2(g) \rightarrow BeCl_2(s)$$
$$Be(s) + Br_2(g) \rightarrow BeBr_2(s)$$

3. Reactivity towards hydrogen- Except beryllium, all alkaline earth metals reacts with hydrogen to form hydrides.

4. Reactivity towards acids- The alkaline earth metals readily react with acids giving out dihydrogen.

$$M + 2HCl \rightarrow MCl_2 + H_2$$

5. Reducing nature- Alkaline earth metals are strong reducing agents. They possess large negative reduction potentials. Beryllium though has less negative reduction potential but it is still reducing in nature due to large hydration energy of beryllium ion (Be^{2+}).

6. Solutions in liquid ammonia- The alkaline earth metals dissolve in liquid ammonia to give blue black color due to the formation of ammoniated ions.

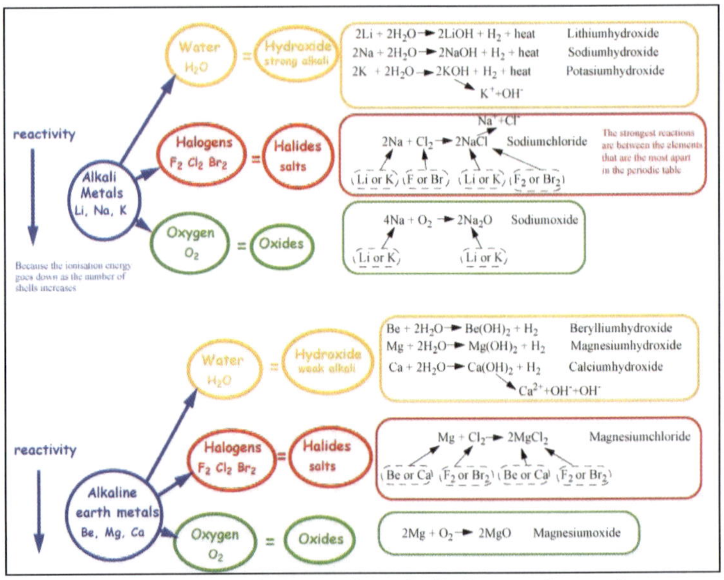

Comparison of alkali metals and alkaline earth metals.

The Group 13 Elements

Group 13 Elements are also called Boron group elements that consist of six chemical elements. The elements are boron (B), aluminum (Al), gallium (Ga), indium (In), thallium (Tl), and nihonium

(Nh). They are characterized as a group by having three electrons in the outermost parts of their atomic structure. Boron, the lightest of these elements, is a metalloid. Aluminum, gallium, indium, and thallium are silvery white metals. Nihonium has only been produced as individual atoms in particle accelerators.

None of these elements was known in a pure state before modern chemistry isolated them. Very soon after a method had been found to produce it in commercial quantities, aluminum revolutionized industry. With further development of science and technology, specifically in nanotechnology, boron gained significant attention in industrial sectors as well. The other members of the group still have little commercial value. Some of the compounds of boron and aluminum, however, are indispensable in modern technology and have been widely used in many parts of the world throughout recorded history.

General Properties of the Group

Ionization Energies

Every element in the boron group has three electrons in its outermost shell (so-called valence electrons), and for each element there is a sharp jump in the amount of energy required to remove the fourth electron, reflecting the fact that this electron must be removed from an inner shell. Consequently, the elements of the group have maximum oxidation numbers of three, corresponding to loss of the first three electrons, and form ions with three positive charges.

The apparently erratic way in which ionization energies vary among the elements of the group is due to the presence of the filled inner *d* orbitals in gallium, indium, and thallium, and the *f*orbital in thallium, which do not shield the outermost electrons from the pull of the nuclear charge as efficiently as do the inner *s* and *p* electrons. In Groups 1 and 2 (Ia and IIa), in contrast to the boron group, outer shell (always referred to as n) electrons are shielded in every case by a constant inner set of electrons, in the $(n-1)s^2(n-1)p^6$ orbitals, and the ionization energies of these Group-1 and Group-2 elements decrease smoothly down the group. The ionization energies of gallium, indium, and thallium are thus higher than expected from their Group 2 counterparts because their outer electrons, being poorly shielded by the inner *d* and *f*electrons, are more strongly bound to the nucleus. This shielding effect also makes the atoms of gallium, indium, and thallium smaller than the atoms of their Group 1 and 2 neighbours by causing the outer electrons to be pulled closer toward the nucleus.

The M^{3+} state for gallium, indium, and thallium is energetically less favourable than Al^{3+} because the high ionization energies of these three elements cannot always be balanced by the crystal energies of possible reaction products. For example, of the simple, anhydrous compounds of thallium in its +3 oxidation state, only the trifluoride, TlF_3, is ionic. For the group as a whole, therefore, the M^{3+} ionic state is the exception rather than the rule. More commonly the elements of the group form covalent bonds and achieve an oxidation state of three by promoting one electron from the *s* orbital in the outer shell (designated n*s* orbital) to an n*p*orbital, the shift permitting the formation of hybrid, or combination, orbitals (of the variety designated as sp^2). Increasingly down the group there is a tendency toward the formation of M^+ ions, and at thallium the +1 oxidation state is the more stable one. The basicity (a property of metals) of the elements also increases in proceeding down the group, as shown by the oxidesthey form: boric oxide (formula B_2O_3) is acidic;

the next three oxides, of aluminum, gallium, and indium (formulas Al_2O_3, Ga_2O_3, and In_2O_3) are either acidic or basic depending on the environment (a property called amphoterism); and thallic oxide (Tl_2O_3) is wholly basic.

Some Properties of the Boron Group Elements

The table gives a list of some properties of the boron group elements.

Some properties of the boron group elements					
	Boron	Aluminum	Gallium	Indium	Thallium
Atomic number	5	13	31	49	81
Atomic weight	10.811	26.982	69.723	114.818	204.383
Color of element	brown	silver-white	gray-blue	silver-white	blue-white
Melting point (°C)	2,075	660.32	26.76	156.6	304
Boiling point (°C)	4,000	2,519	2,204	2,072	1,473
Density: solid (grams per cubic centimetre at 20 °C)	2.34	2.699	5.904*	7.31	11.85
Density: liquid (grams per millilitre)	2.37	2.375	6.095	7.02	11.22
Valence	3	3	3	3, 1	3, 1
Mass number of most common isotopes (terrestrial abundance, percent)	10 (19.92), 11 (80.12)	27 (100)	69 (60.108), 71 (39.892)	113 (4.29), 115 (95.71)	203 (29.52), 205 (70.48)
Radioactive isotopes (mass numbers)	7–9, 12–19	21–26, 28–41	60–68, 70, 72–86	97–112, 114–135	176–202, 204, 206–212
Color imparted to flame	green	Colorless	violet	blue	green
Color of ions in solution: +3	—	Colorless	Colorless	Colorless	Colorless
Color of ions in solution: +1	—	—	—	—	Colorless
Heat of fusion (calories per mole/kilojoules per mole)	12,000 (50)	2,560 (10.7)	1,340 (5.59)	779 (3.26)	1,000 (42)
Specific heat (joules per gram Kelvin)	1.026	0.897	0.373	0.233	0.129
Electrical resistivity at 20–25 °C (microhm-centimetres)	>1012	2.7	14	8	15
Hardness (Mohs' scale)	9.3	2.75	1.5	1.2	1.2
Crystal structure at 20 °C	alpha-rhombohedral, beta-rhombohedral, tetragonal	face-centred cubic	orthorhombic	face-centred tetragonal	hexagonal close-packed
Radius: atomic (angstroms)	0.87	1.18	1.36	1.56	1.56
Radius: ionic (angstroms)	0.41	0.68	0.76	0.94	1.03
Ionization energy (electron volts): first	800.6	577.5	578.8	558.3	589.4
Ionization energy (electron volts): second	2,427.10	1,816.70	1,979.30	1,820.70	1,971
Ionization energy (electron volts): third	3,659.70	2,744.80	2,963	2,704	2,878
Ionization energy (electron volts): fourth	25,025.80	11,577	6,180	5,210	—

Oxidation potential for oxidation from the 0 to +3 oxidation state at 25 °C (volts)	—	1.68	0.53	0.34	−1.25
Electronegativity (Pauling)	2.04	1.61	1.81	1.78	1.62
Density at 25 °C					

Compounds of the Boron Group Elements

Salts of M^{2+} Ions

The ionization energies suggest that the formation of salts of the M^{2+} ions might be feasible. At first glance, such appears to be the case, since gallium compounds with the formula GaX_2 (X representing chlorine, bromine, or iodine) can be made, and similar cases occur with the other metals of this group. Such compounds, however, are generally found to be of mixed oxidation state; that is, they contain metal atoms in both the one and the three oxidation states, a condition symbolized as $M^+(M^{3+}X_4)^-$. The nearest approach to M^{2+} derivatives occurs in gallium sulfide, selenite, and telluride, which are made by heating gallium with stoichiometric amounts of sulfur, selenium, and tellurium, respectively. Studies of the structure of these compounds by X-ray methods show that they contain $(Ga\text{-}Ga)^{4+}$ units arranged in a layer like lattice; the coupling of the gallium atoms in such a manner pairs the electrons available for the bonds and thereby explains the diamagnetism of the compounds (diamagnetism is a property associated with paired electrons).

The large amount of energy required to remove three electrons completely from a boron atom makes the formation of salts containing the bare B^{3+} cation impossible; even water of hydration associated with such ions would be too highly deformed to be stable, and hence the aquated ion $B^{3+}(aq)$ is unknown. Much less energy is required to promote electrons from 2*s*orbitals into 2*p* orbitals in boron atoms, with the result that boron compounds are always covalent. The boron orbitals are hybridized to either the sp^2 (when boron forms bonds with three other atoms, for example, in borazine) or the sp^3 (when boron forms bonds with four atoms, as in metal borohydrides) configuration.

Hydrated Ions in the +3 Oxidation State

Although simple M^{3+} cations are uncommon in anhydrous compounds of the boron group elements, the hydrated (combined with water) triply charged ions of aluminum, gallium, indium, and thallium are well known in water solution. Nuclear magnetic resonance studies reveal that six water molecules are held strongly by these positive ions in solution, and their salts often can be crystallized from solution combined with six water molecules. The high charge on the central cation of such hydrates induces the ionization of protons, or hydrogennuclei, on the coordinated water molecules and thereby leads to the formation of basic salts.

This reaction (called hydrolysis) is represented in the following equations:

$$M(H_2O)_n^{3+} \rightleftharpoons (HO)M(H_2O)_{n-1}^{2+} + H^+(aq) \rightleftharpoons$$

$$(HO)_2M(H_2O)_{n-2}^{+} + 2H^+(aq) \rightleftharpoons M(OH)_3 + 3H^+(aq)$$

In which, as before, M represents an ion of one of the boron group elements, n is the number of water molecules joined to it, (HO)M represents a hydroxide group joined to the metal ion, and H^+(aq) is a hydrated hydrogen ion. In these and other equations, the arrows pointing in two directions indicate that the chemical reactions can proceed both ways, depending on the reaction conditions. When acid is added to such aqueous solutions, it depresses the hydrolytic processes by reversing the reactions. At high acid concentrations, however, complex anions (negative ions) are sometimes formed, especially with the aqueous hydrogen halides. The following equation illustrates this: Ga^{3+}(aq) + HX (conc.) → GaX_4^-, X being chlorine, bromine, or iodine. Intermediate complex ions, MX^{2+} and MX_2^+, can be detected in several cases.

Trihalides

The electrical conductivity of solid aluminum trichloride (formula $AlCl_3$), in which each aluminum ion has three positive charges, increases rapidly as the temperature is elevated toward the melting point, at which the conductivity suddenly falls to zero. This phenomenon occurs because the aluminum and chloride ions form an ionic lattice that partially conducts electricity, but upon melting, the compound changes to the electrically nonconducting covalent state. The explanation is that the distribution of energy in the liquid state is insufficient to compensate for the ionization energy required to separate the Al^{3+} and Cl^- ions, and these then acquire covalent bonds. The liquid consists of double or dimeric molecules with the formula Al_2Cl_6, which may be represented in the following manner that shows a molecule with the position of its atoms in three dimensions; the solid lines are in the plane of the screen, the dotted lines are behind the screen, and the shaded lines indicate that they extend toward the viewer:

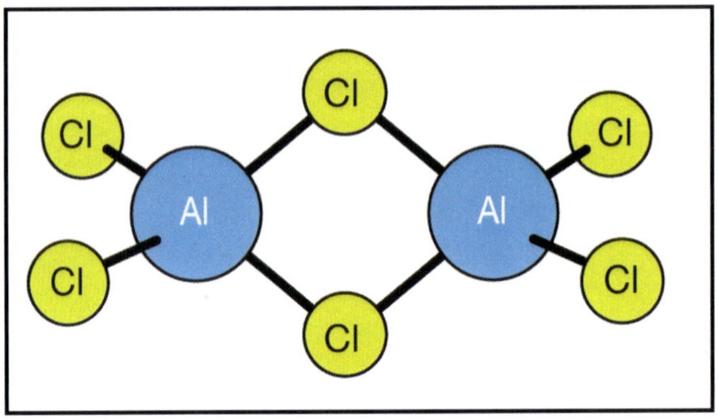

Boron group element: Dimeric molecules with the formula Al_2Cl_6.

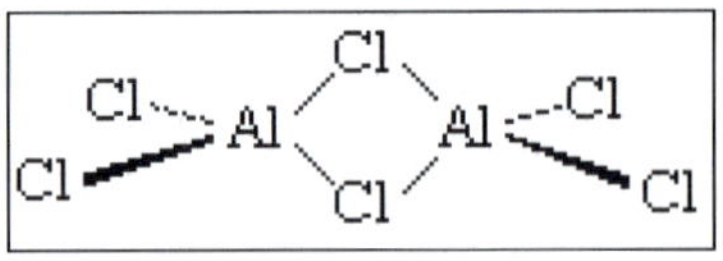

The delicate energy balance between ionic and covalent bonding for aluminum in the +3 oxidation state can be appreciated when it is realized that whereas solid aluminum trifluoride, formula AlF_3, is ionic like the chloride, aluminum tribromide forms molecular crystalscontaining dimers, with the formula Al_2Br_6.

In contrast to the dimers, the single, or monomeric, trihalides of the boron group elements have trigonal planar structures. If M is the metal and X is any halogen, the arrangement of the atoms can be sketched as follows:

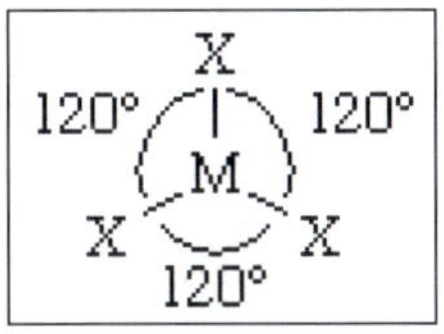

The trihalides of boron have this configuration in all phases, whereas the trihalides of aluminum, gallium, indium, and thallium become monomeric only on being heated in the gas phase. In MX_3 molecules, the central atom M has added three electrons to its own, making only six electrons in the outer shell, although eight are required to achieve the desired inert-gas configuration. These halides, therefore, readily accept two more electrons from many donor molecules (e.g., ethers, alcohols, amines, and phosphines) that carry unshared pairs of electrons.

A typical case, the reaction of gallium tribromide with trimethylamine, is represented in the following equation,

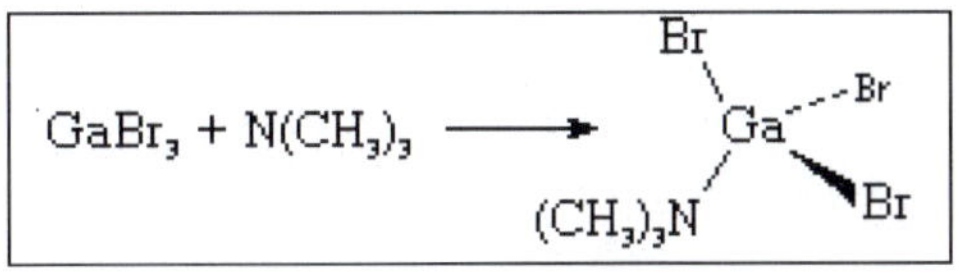

The central gallium atom is coordinated or bonded to three bromine atoms and one nitrogenatom. The electron donor also can be a halide ion, in which case the tetrahedral complex anion, MX_4^- results.

Less-common Compounds

A few compounds are known in which aluminum, gallium, indium, and thallium are coordinated to five or six atoms. These compounds have structures of the following types, M again representing any boron group element, D any donor molecule, and X any halogen (again, the solid lines are bonds in the plane of the screen, the atoms so bonded lying in that plane; the dotted lines lead behind the screen; the shaded lines reach toward the viewer):

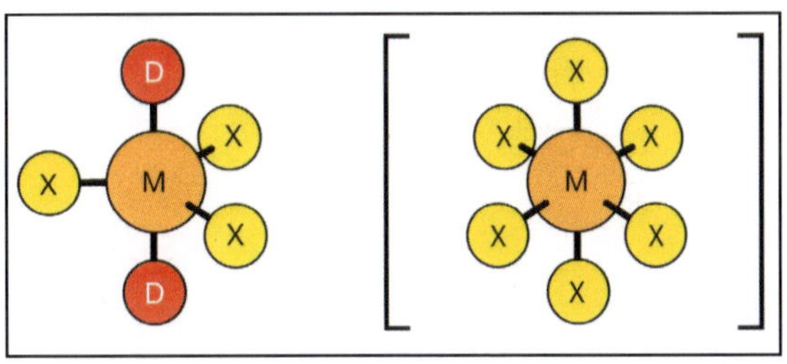

Boron group element: An uncommon compound structure for a boron group element where M represents a boron group element, D is a donor molecule, and X is any halogen.

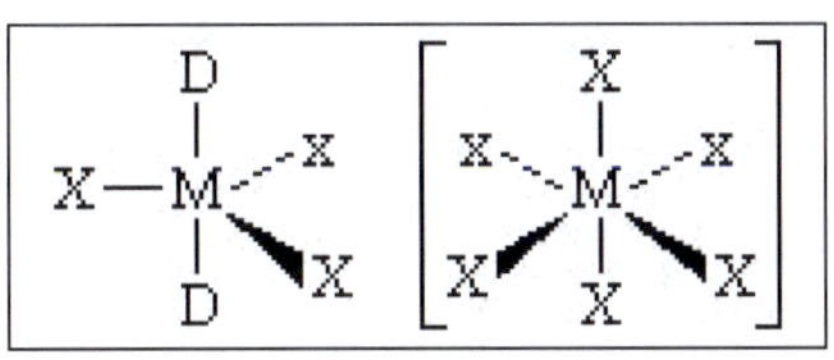

In such compounds it is possible, but by no means certain, that the central element makes use of its vacant n*d* orbitals to increase its oxidation state by way of sp^3d (five-coordination) or sp^3d^2 (six-coordination) hybridization. If the concept of the participation of *d* orbitals in the bonding of these compounds is valid, it would account for the fact that boron, which has no available *d* orbitals, does not form five- and six-coordinate compounds. In many cases, however, spatial requirements also would rule out the possibility of boron's increasing its covalency above four, because the boron atom is so small that no more than four atoms can be arranged around it.

In the gas phase at high temperature, all the boron group elements form diatomic halides MX, either by dissociation of the trihalides or, more commonly, by reduction of the trihalides with the free element, as in the following equations for two such reactions:

$$\underset{\text{boron}}{2B} + \underset{\text{boron trifluoride}}{BF_3} \xrightarrow{2{,}000^{o}} \underset{\text{boron monofluoride}}{3BF}$$

$$\underset{\text{aluminum}}{2Al} + \underset{\text{aluminum trichloride}}{AlCl_3} \xrightarrow{800^{o}\text{-}1{,}000^{o}} \underset{\text{aluminum monochloride}}{3AlCl}$$

Most of these monohalides, especially those of boron, aluminum, and gallium, are unstable in the solid state under normal conditions. They exist only at high temperatures as gases. All are covalently bonded, except thallium fluoride, which exists as the ion pair, $Tl^+ F^-$.

Boron and aluminum can also form organoboron and organoaluminum compounds, where the group 13 element is directly bonded to a carbon atom. These types of compounds, acting mostly as catalysts, are integral to a variety of organic syntheses. Certain organoboron compounds are useful in medicine.

Thallium is the only element that forms a stable ion having an $(n\text{-}1)d^{10}ns^2$ outer electronic configuration. There is, therefore, no ion to which direct comparisons with the singly charged thallium ion, Tl^+, might be made.

Diagonal Relationship of Beryllium and Aluminum

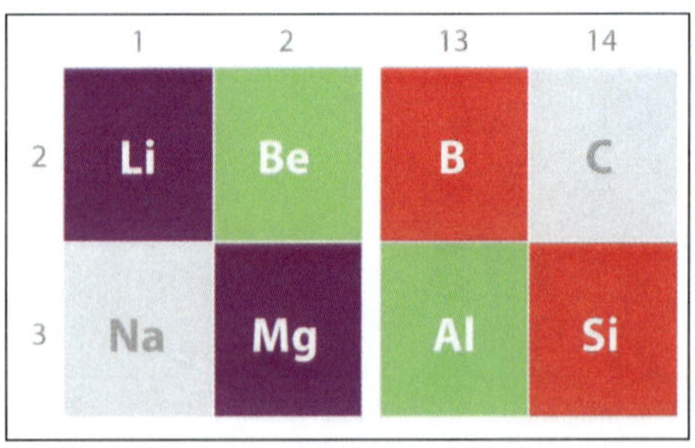

Diagonal relationships in the periodic table.

Beryllium and aluminum share a great deal of similitudes and this is known as diagonal relationship. Both Be_2^+ and Al_3^+ in hydrated condition, deliver $[Be(H_2O)_4]_2^+$ and $Al(H_2O)_{63}$ separately. At the point when reacted with water, both compounds create hydronium particles, making them somewhat acidic. Another similitude amongst aluminum and beryllium is that they are amphoteric, and their hydroxides are exceptionally basic. Both metals additionally react with oxygen to deliver oxide coatings fit for shielding different metals from corrosion. Both metals additionally react with halides that can go about as Lewis acids.

Applications of Boron Group

- A regular utilization of Boron is in fiberglass. There has been a fast development in the market for borosilicate glass; most remarkable among its uncommon qualities is a much more noteworthy resistance to thermal expansion than normal glass.
- Another commercially extending utilization of boron and its subordinates is in ceramics. A few boron compounds, particularly the oxides, have exceptional and profitable properties that have prompted to their substitution for different materials that are less helpful. Boron might be found in ceramic panhandles, pots, plates, vases, due to its insulating properties.
- Aluminium is frequently utilized as a part of construction materials, electrical gadgets, particularly as a transmitter in links, and in apparatuses and vessels for cooking and safeguarding food. Aluminum's absence of reactivity with food items makes it especially helpful for canning.
- Aluminium is a part of alloys utilized for making lightweight bodies for flying machine. Cars additionally fuse aluminum in their structure and body some of the time, and there are comparative applications in military gear and vehicles.
- Gallium arsenide has been utilized as a part of semiconductors, enhancers, solar cells (for instance in satellites) and passage diodes for FM transmitter circuits.
- Gallium amalgams are utilized for the most part for dental purposes. Gallium ammonium chloride is utilized for the leads in transistors. A noteworthy use of gallium is in LED lighting.
- Indium might be found on platings, phosphors, bearing, display gadgets, warm reflectors, and atomic control bars. Indium tin oxide has found an extensive variety of utilizations, including glass coatings, solar panels, road lights, electrophoretic displays (EPDs), plasma display boards (PDPs), electroluminescent light displays (ELDs), electrochemical displays (ECs), sodium lights, field emission displays (FEDs), windshield glass and cathode ray tubes.

Property	Bron	Aluminum	Gallium	Indium	Thallium
Atomic symbol	B	AI	Ga	In	TI
Atomic number	5	13	31	49	81
Atomic mass (amu)	10.81	26.98	69.72	114.82	204.38
Valence electron configuration†	$2s^2 2p^1$	$3s^2 3p^1$	$4s^2 4p^1$	$5s^2 5p^1$	$6s^2 sp^1$
Melting point/ boiling point(°C)	2075/4000	660/2519	29.7/2204	156.6/2072	304/1473
Density (g/cm^3) at 25 °C	2.34	2.70	5.91	7.31	11.8

Atomic radius (pm)	87	118	136	156	156
First ionization energy (kJ/mol)	801	578	579	558	589
Most common oxidation state	+3	+3	+3	+3	+1
Ionic radius (pm)[‡]	-25	54	62	80	162
Electron affinity (kJ/mol)	-27	-42	-40	-39	-37
Electronegativity	2.0	1.6	1.8	1.8	1.8
Standard reduction potential (E°,V)	-0.87	-1.66	-0.55	-0.34	+0.741 of M^{3+} (aq)
Product of reaction with O_2	B^2O^3	AI^2O^3	Ga^2O^3	In^2O^3	TI^2O
Type of oxide	acidic	amphoteric	amphoteric	amphoteric	basic
Product of reaction with N_2	BN	AIN	GaN	InN	none
Product of reaction with X_2[δ]	BX_3	AI_2O_3	Ga_2X_6	In_2X_6	TIX

[#]This is the name used in the United States; the rest of the world inserts an extra i and calls it aluminium.

[†]The configuration shown does not include filled d and f subshells.

[‡]The values cited are for six-coordinate ions in the most common oxidation state, except for AI^{3+}, for which the value for the four-coordinate ion is given. The B^{3+} ion is not a known species; the radius cited is an estimated four-coordinate value.

[δ]X is CI,Br, or I. Reaction with F_2 gives the trifluorides (MF_3) for all group 13 elements.

The Group 14 Elements

The group 14 elements are the second group in the p-block of the periodic table. It is also called the carbon group.

The members of this group are:

- Carbon (C)
- Silicon (Si)
- Germanium (Ge)
- Tin (Sn)
- Lead (Pb)
- Flerovium (Fl)

Electronic Configuration of Group 14 Elements

The group 14 elements have a general electronic configuration of ns^2np^2. These elements have 2 electrons in the outermost p orbitals. The electronic configuration of these elements is shown below:

Group 14 Elements			
Period	Element	Atomic No.	Electronic Configuration
2nd	Carbon (C)	6	$[He]_2s^2\ 2p^2$
3rd	Silicon (Si)	14	$[Ne]_3s^2\ 3p^2$
4th	Germanium (Ge)	32	$[Ar]_3d10\ 4s^2\ 4p^2$
5th	Tin (Sn)	50	$[Kr]_4d10\ 5s^2\ 5p^2$
6th	Lead (Pb)	82	$[Xe]_4f_{14}\ 5d10\ 6s^2\ 6p^2$

As all the elements in group 14 have 4 electrons in the outermost shell, the valency of group 14 elements is 4. They use these electrons in the bond formation in order to obtain octet configuration.

Oxidation States and Inert Pair Effect of Group 14 Elements

The general oxidation states exhibited by the group 14 elements are +4, and +2. As we go down the group, the tendency to form +2 ion increases. This is due to inert pair effect. This effect is exhibited by p-block elements. This can be explained using the inert pair effect. It is the non-participation of the s-orbital during bonding due to the poor shielding of the intervening electrons.

For elements like Sn and Pb, d and f orbitals are filled with electrons. Since the shielding ability of d and f orbitals are very poor, the nuclear charge that seeps through attracts the s orbital closer to the nucleus. This makes the s orbital reluctant to bond, thereby only the p electrons involved in bonding.

Therefore, Pb4+ is a very good oxidizing agent.

Anomalous Behavior of Carbon

Carbon exhibits different behavior from the rest of the group due to:

- Small Size;
- High Electronegativity;
- High Ionization Enthalpy;
- Absence of d-orbital in the Valance Shell.

Chemical Properties of Group 14 Elements

1. Covalent Radii:

   The radii of group 14 elements are lesser than that of group 13 elements. This can be explained by the increase in the effective nuclear charge.

   The increase in the radii from C to Si is considerable, after which the increase in the radii is less. This can be attributed to the poor shielding of d and f orbitals, which increase the effective nuclear charge, thereby making the radii small.

2. Ionization Enthalpy:

   The ionization energy of group 14 elements is greater than that of group 13 elements. This can be attributed to size.

   Down the group, the Ionization Enthalpy decreases. There is a sharp decrease from C to Si, after which the decrease is nominal.

   The order is as follows, C > Si > Ge > Pb >Sn

   Here Pb has a greater Ionization Enthalpy than Sn due to ineffective shielding of d and f orbitals.

Physical Properties

1. Metallic Character:

   Group 14 elements are less electropositive than group 13 owing to their small size and high ionization enthalpy. Down the group, the metallic character increases. C and Si are non-metals, Ge a metalloid, and Sn and Pb are soft metals with low melting points.

2. Melting and Boiling Points:

   The melting and boiling point of carbon, silicon, and germanium is significantly high because they have a very stable solid structure. Sn and Pb have a lower melting point because only two bonds are formed instead of four, due to inert pair effect.

   The melting point of carbon is extremely high. All the elements of group-14 possess diamond-type lattice structure which is highly stable in nature. The process of melting results in the breakage of these highly stable lattice structures.

   Down the group, the melting point decreases as the M-M bonds are reduced as the size of the atoms increases. Since, Tin and lead are metals therefore, the melting points of these elements are much lower.

Four Covalent Compounds

Four covalent compounds are compounds in which the four electrons in the valence shell play an active role in bonding. Most of the group 14 element possesses this property.

Compounds Formed by Group 14 Elements

Hydrides of Group 14

All of the elements of group 14 form hydrides. Carbon forms hydrides extensively due to their ability to catenate. The hydrides of carbon are categorized as below:

- Alkanes (paraffin's): General formula: C_nH_{2n+2}.
- Alkenes (olefins): General formula: CnH_2n.

- Alkynes (Acetylenes): General formula: C_nH_{2n-2}.
- Aromatic compounds.
- Silicon forms hydrides having the general formula Si_nH_{2n+2} and are called silanes.
- Germanium forms hydrides of the form Ge_nH_{2n+2}, where n_{max}=5 and are called germanes.
- Tin forms SnH_4 and are called stannane. It is much less stable.

Oxides of Group 14

Group 14 elements form oxides of the type MO and MO_2. Lead also form an oxide Pb_3O_4 which is a mixed oxide of PbO and PbO_2. Among the monoxides, CO is neutral GeO is basic while SnO and PbO are amphoteric.

In CO_2, C is sp hybridised - It is different from SiO_2 in which Si is sp^3 hybridised. In SiO_2, each O atom is bonded to two Si bonds. This gives rise to a three-dimensional structure for SiO_2. This also attests to the high melting point of SiO_2. Among the dioxides, the acidic character decreases down the group. CO_2 is the most acidic and PbO_2 being the most basic amongst the dioxides.

Halides of Group 14

They form tetrahalides of the form MX_4. The central atom is sp^3 hybridized and assumes a tetrahedral shape.

Elements below C, have empty d-orbitals, with which they can exhibit back bonding with the halogens (pπ-dπ). Carbon does not form dihalides. The dihalides are sp^2 hybridized and have a bent shape.

Electronic Configuration of the Carbon Family

Electronic configuration of an atom is nothing but an illustration of the layout of electrons distributed among the sub-shells and orbitals. By this configuration of electrons, we can understand the various physical and chemical properties of the elements. The chemistry behind the elements can be determined by studying the number of valence electrons in the outermost shells.

Before understanding the electronic configuration of elements, we must understand the rules for assigning the electrons into the orbitals. There are many principles that help us in doing so. These include Pauli's exclusion principle, Hund'srule of maximum multiplicity and Aufbau principle.

Electrons fill the orbitals in such a way that the energy of the atom is at the minimum. Hence, the electrons of an element fill the energy levels in an increasing order as per the Aufbau principle. Pauli defined a set of unique quantum numbers for each electron. Pauli Exclusion Principle states that all the four quantum numbers for any two electrons in an atom can never be same.

As per Hund's rule, the pairing of electrons in an orbital takes place only when all the sub-shells have one electron each. The general electronic configuration of these group 14 elements is ns^2np^2.

These elements have 2 electrons in the outermost p orbitals. The electronic configuration of group 14 elements is as follows:

Period	Element	Symbol	Atomic Number	Electronic Configuration
2	Carbon	C	6	$[He]2s^2 2p^2$
3	Silicon	Si	14	$[Ne]3s^2 3p^2$
4	Germanium	Ge	32	$[Ar]3d^{10} 4s^2 4p^2$
5	Tin	Sn	50	$[Kr]4d^{10} 5s^2 5p^2$
6	Lead	Pb	82	$[Xe]4f^{14} 5d^{10} 6s^2 6p^2$

As all the elements in group 14 have 4 electrons in the outermost shell, the valency of Group-14 elements is 4. They use these electrons in the bond formation in order to obtain octet configuration.

The Group 15 Elements

The p-block elements are put to the right-hand side of the periodic table in groups from 13 to 18. In the iotas of p-block elements, the separating electron enters the valence p subshell. In this manner, in these elements, the np subshell is filled step by step.

The general valence shell electronic setup of group fifteen elements is ns^2, np^{1-6}. The electronic design of helium is $1s^2$. It has no orbitals. However, it is a p-block element since it takes the physical and chemical properties after that of other p-block elements of the eighteenth group. P-block elements are generally non-metals, while the remaining are metalloids and metals.

Occurrence

Group 15 elements include nitrogen, phosphorus, arsenic, antimony and bismuth. Nitrogen is the real constituent of the air and records for 78% of it by volume. It is the primary member of this group and happens in a free state as a diatomic gas, N_2.

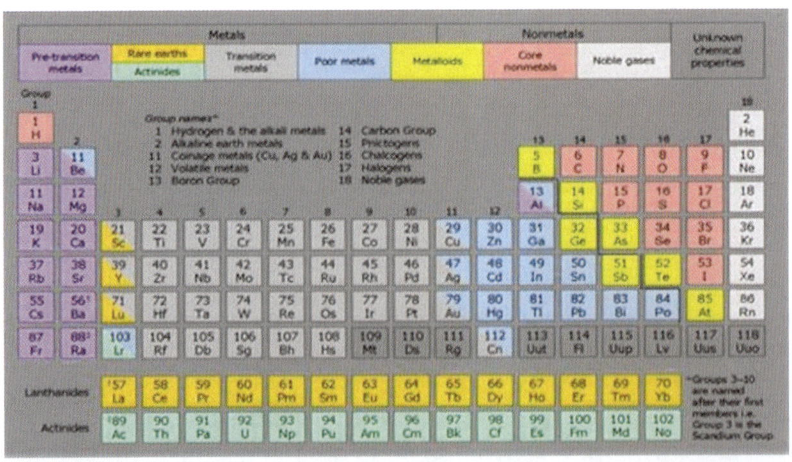

Phosphorus is a fundamental constituent of animal and plant matter. Phosphate groups are constituents of nucleic acids, that is, DNA and RNA. Around 60% of bones and teeth are made out of phosphates. Phosphoproteins are available in egg yolk, milk, and bone marrow. The rest of the elements of the group, that is, arsenic, antimony, and bismuth, mostly happen as sulfides. For example, Stibnite, Arsenopyrite, and bismuth glance.

Trends in Group 15 Elements

Let us focus at some of the trends of atomic properties of these elements.

Atomic Radii

Moving down the group, the ionic radii, and atomic radii increases. This is because of the expansion of another main energy level in each progressive element.

Ionization Enthalpy

These elements demonstrate higher values of ionization enthalpy when contrasted with group 14 elements. As we move down the group, the ionization enthalpy values keep on decreasing. This is because of the progressive increase in the size of the nucleus.

Electronegativity

Electronegativity is the inclination of a particle to pull in a shared pair of electrons more towards itself. The electronegativity decreases gradually on moving down the group. The reason behind this is the increase in atomic radius.

Physical Properties

Physical properties incorporate physical state, boiling and melting points, metallic character, allotropy, and density. Nitrogen is a diatomic gas, while the rest of the elements are solids in nature.

Moving down a group, metallic character increases. On the other hand, the ionization enthalpy of the elements decreases due to an increase in their nuclear size.

Trends in Melting and Boiling Points

The melting point increments from nitrogen to arsenic because of the continuous increment in nuclear size. The low melting point of nitrogen is because of its discrete diatomic particles.

In spite of the fact that the nuclear size increments from arsenic to antimony, there is a decrease in their melting points. Despite the fact that antimony has a layered structure, it has a low melting point than arsenic on account of the generally free pressing of particles. Furthermore, the melting point of bismuth is not as much as antimony because of the packing of atoms loosely by metallic holding. Then again, the boiling point steps by step increments from nitrogen to bismuth.

The density of these elements increases from nitrogen to bismuth.

Allotropy

All group fifteen elements, aside from bismuth, indicate allotropy. Nitrogen is found in two allotropic structures, that is, alpha nitrogen and beta nitrogen. Phosphorus exists in numerous allotropic structures. Of these, the two critical allotropic structures are red phosphorus and white phosphorus.

Arsenic exists in three essential allotropic structures – black, grey, and yellow. Antimony additionally has three essential allotropic structures, to be specific, yellow, metallic and explosive.

Electronic Configurations

Similarities in Orbital Arrangement

In the periodic table, each of the nitrogen group elements occupies the fifth position among the main group elements of its period, a position designated 15. In terms of the electronic configuration of its atoms, each nitrogen group element possesses an outermost shell of five electrons. In each case, these occupy an outer *s* orbital completely (with two electrons) and contribute one electron to each of the three outer *p* orbitals (the orbitals being electron regions within the atom and the letter designations, *s, p, d*, and *f*, being used to designate different classes of orbital). The arrangement of outer electrons in the atoms of the nitrogen elements thus provides three half-filled outer orbitals that, by interaction with half-filled orbitals of the atoms of other elements, can form three covalent bonds. The other atoms may attract the shared electrons either more or less strongly than do the nitrogen group atoms; therefore the latter may acquire either positive or negative charges and exist in oxidation states of +3 or −3 in their compounds. In this respect, the nitrogen elements are alike.

Another similarity among the nitrogen elements is the existence of an unshared, or lone, pair of electrons, which remains after the three covalent bonds, or their equivalent, have been formed. This lone pair permits the molecule to act as an electron pair donor in the formation of molecular addition compounds and complexes. The availability of the lone pair depends upon various factors, such as the relative size of the atom, its partial charge in the molecule, the spatial characteristics of other groups in the molecule, and the as-yet poorly understood phenomenon called the "inert pair effect." This effect consists of a tendency for the paired selectors in the outermost shell of the heavier atoms of a major group to remain chemically unreactive. Because of it, the electron pair-donating ability of the nitrogen group elements is not uniform throughout the group; it is probably greatest with nitrogen, less with the intermediate elements, and nonexistent with bismuth.

Variations in Bonding Capacity

Significant differences in electronic configurations also occur among the elements of the nitrogen group with respect both to the underlying shell and to the outer *d* orbitals. Since the latter first appear with the third period of the table, they are present in all elements of the group but nitrogen. The possibility of utilizing these outer *d* orbitals for bonding thus exists for phosphorus, arsenic, antimony, and bismuth, but not for nitrogen.

There are three principal ways in which the outer *d* orbitals can be used to increase the number of bonds or expand the valence octet. One is by providing a space to which one of the *s* electrons can be promoted. This creates two additional half-filled orbitals (one *d* and one's orbital), and it therefore generates the capacity to form two additional covalent bonds. This is exemplified by the production of phosphorus pentafluoride, PF_5, by further fluorination of the trifluoride, PF_3. Such promotion appears to be greatly assisted by the increase in outer *d*-orbital stability that results from the withdrawal of part of the screening electron and the attendant increase of the effective nuclear charge of the central atom. In PF_5, for example, the fluorine atoms, being much more electronegative than the phosphorus atom, draw away a portion of the phosphorus electrons, leaving the outer *d*orbitals more exposed to the phosphorus nucleus and therefore more stable.

A second way in which the outer *d* orbitals can become involved in the bonding is by their becoming sufficiently stable to attract a lone pair of electrons from a donor. For example, PF_5 can serve as an electron pair acceptor through an outer *d* orbital to coordinate a fluoride ion donor and form the complex ion PF_6^-.

A third way of involving *d* orbitals in bonding is for them to become partially occupied in accommodating lone-pair electrons from another atom, which is already attached by a single bond, thereby strengthening the bond. The phosphorus oxyhalides, of general formula POX_3, appear to be examples of this; their phosphorus–oxygen bonds are observed to be shorter and stronger than expected for ordinary single bonds.

The +5 Oxidation State

It thus is possible for an atom of phosphorus, arsenic, antimony, or bismuth to expand its valence octet to form five covalent bonds and one additional coordinate covalent bond. This is not possible for nitrogen, which exhibits a maximum coordination number of four: three single covalent bonds and a coordinate covalent bond with nitrogen acting as donor (through its lone pair). Nevertheless, the +5 oxidation state is formally applicable to nitrogen, so that nitrogen, phosphorus, arsenic, antimony, and bismuth can all be found in this state. When compounds in the +5 oxidation state are studied, however, it is observed that their properties do not exhibit a uniform trend within the group. Rather, a certain degree of alternation is observed, the +5 states of nitrogen, arsenic, and bismuth appearing less stable and more strongly oxidizing than the corresponding states of phosphorus and antimony. In part this alternation may find explanation in the electronic differences among the atoms with respect to their underlying shells. The number of electrons in the shell just below the outermost level, is two for nitrogen, eight for phosphorus, and 18 for arsenic, antimony, and bismuth.

Increasing the nuclear charge by 18 from phosphorus to arsenic may be accompanied by incomplete shielding of this extra charge by the ten 3d electrons also added. This would imply smaller size and a greater electronegativity for arsenic than for phosphorus and thus a greater similarity between the phosphorus and antimony atoms. This subject, however, is still controversial, and the widely used scale of electronegativities devised by Linus Pauling fails to make this distinction.

An interesting anomaly is presented by the fact that nitrogen as a free element is in the form of gaseous diatomic molecules, while the elements immediately preceding it in its period of the table are solids, as are the other elements in its group. In surveying the elements of the second period, the most obvious difference in atomic structure found on reaching nitrogen is the appearance for the first time in compounds of the element of a lone pair of electrons not used in bonding with other atoms. Calculations suggest that the presence of this lone pair of electrons is associated with a considerable weakening of nitrogen to nitrogen single bonds in compounds where these bonds occur. In the diatomic nitrogen molecule, however, the bonding is of a different variety—triple bonds being found between the atoms. It is thought that the triple bond is unaffected (unweakened) by the lone pairs of electrons on the nitrogen atoms, and this is assumed to be the reason why nitrogen "prefers" to exist as triply bonded gaseous diatomic molecules rather than as a condensed singly bonded solid polymer.

The same effect might be expected to be operable with the other elements of the nitrogen group, all of which also contain lone electron pairs in their outermost shells. Further calculations disclose, however, that the bond-weakening effect of the lone pair is far less pronounced with these elements than it is with nitrogen. As a result, with these elements, single bonds are favoured over multiple bonds, and the diatomic state of the molecules is not the preferred form.

Relative Electronegativities

It might also be expected that the weakening effect of the lone pair would be observed in compounds of the nitrogen group elements. The picture is more complicated here because the bonds under discussion are formed between different types of atoms. Since different elements differ in electronegativity, bonds between the atoms of different elements are inevitably polar. For purposes of discussion it can be assumed that polar bonds consist of blends of nonpolar covalent bonds and completely polar, ionic bonds. It can then be shown that a relatively small amount of ionic character will contribute a disproportionate share to the overall bond strength. Since the weakening effect of the lone pair is felt only on the covalent portion of the polar bond, rather than on the ionic portion, the less polar bonds will exhibit the greater lone-pair weakening effects.

Comparison of Nitrogen Group Elements

These considerations become important in comparing the chemical behavior of the nitrogen group elements. The electronegativity of nitrogen itself, although lower than that of oxygen, is substantially higher than that of any of the other elements of this group. Bonds between nitrogen and oxygen, therefore, will be considerably less polar than those between oxygen and phosphorus, or oxygen and arsenic, antimony, or bismuth. Consequently, for this reason alone, the covalent contribution to the nitrogen–oxygen bond energy will be relatively more important than is the case with the bonds between oxygen and the heavier elements of the group. Thus, single-bond weakening by the lone pair—and a corresponding tendency toward bond multiplicity—is likely to be much greater with oxides of nitrogen than with oxides of the heavier nitrogen group elements.

The Group 16 Elements

The elements oxygen, sulfur, selenium, tellurium and polonium comprise the 16th vertical column or VIA group elements in the currently used long type of periodic table.

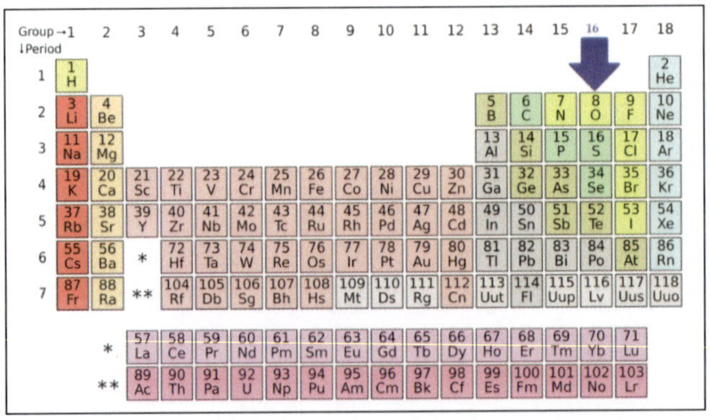

The initial four elements of the group are together termed as chalcogens or ore-forming elements. This is because an extensive number of metal ores are found in the earth's crust as sulfides or oxides. Oxygen is the most plenteous element that is accessible in nature. It shapes 20.946% of air by volume and 46.6% of the world's mass generally as silicates and different compounds like carbonates, oxides, and sulfates.

The vast majority of the oxygen in the air is delivered by photosynthesis in plants. It additionally occurs as ozone. Sulfur is the sixteenth most inexhaustible element. Sulfur in its combined state is found in ores.

Electronic Configuration of Group 16 Elements

Group 16 elements have 6 electrons in their valence shell and their general electronic configuration is ns^2np^4.

Element	Electronic Configuration
Oxygen	$[He]\ 2s^2\ 2p^4$
Sulphur	$[Ne]\ 3s^2\ 3p^4$
Selenium	$[Ar]\ 3d^{10}\ 4s^2\ 4p^4$
Tellurium	$[Kr]\ 4d^{10}\ 5s^2\ 5p^4$
Polonium	$[Xe]\ 4f^{14}\ 5d^{10}\ 6s^2\ 6^4$

Atomic, Physical Properties and their Trends

- Atomic and Ionic Radii: The atomic and ionic radius increases as we move from oxygen to polonium.
- Ionization Enthalpy: Ionization enthalpy decreases with increase in the size of the central atom. Therefore, it decreases as we move from oxygen to polonium since the size of the atom increases as we move down.
- Electron Gain Enthalpy: The electron gain enthalpy decreases with increase in the size of the central atom moving down the group. Oxygen molecule has a less negative electron gain enthalpy than sulfur. This is on the grounds that, oxygen, because of its compressed nature encounter more repulsion between the electrons effectively present and the approaching electron.
- Electronegativity: The electronegativity decreases as we move down the group. Therefore, it decreases as we move from oxygen to polonium due to increase in nuclear size.
- Nature of the Group 16 Elements: Oxygen and sulfur are non-metals, selenium and tellurium are metalloids and polonium is a metal under typical conditions. Polonium is a radioactive element.
- Allotropy: Each one of the element of group 16 displays allotropy. Oxygen has two allotropes: Oxygen and Ozone. Sulphur exist as many allotropic forms but only two of them are

stable, which are: Rhombic Sulphur and Monoclinic Sulphur. Selenium and Tellurium are found in both amorphous and crystalline forms.

- The Melting and Boiling Points: As the atomic size increases from oxygen to tellurium, the melting and boiling points also increase. The huge distinction between the melting and boiling points of oxygen and sulfur might be clarified on the premise that oxygen exists as a diatomic atom while sulfur exists as a polyatomic particle.
- Oxidation States: The group 16 elements have a configuration of ns^2 np^4 in their outer shell, they may accomplish noble gas configuration either by the gain of two electrons, framing M^{-2} or by sharing two electrons, in this manner shaping two covalent bonds.

Thus, these elements indicate both negative and positive oxidation states. The regular oxidation states showed by the elements of group 16 incorporate -2, +2, +4, and + 6.

Chemical Properties

The group sixteen elements react with hydrogen to form hydrides of the sort H_2E, where E could be any element- oxygen, sulfur, selenium, tellurium or polonium.

$$\underset{\text{Hydrogen}}{H_2} + \underset{\text{Group16 Element}}{E} \rightarrow \underset{\text{Hydride}}{H_2E}$$

The Physical States of Hydrides of Group 16 Elements

Water is an odourless and Colorless liquid but the hydrides of the various elements of this group are poisonous gases which are Colorless with disagreeable smells.

The boiling point of these hydrides extraordinarily diminishes from water to hydrogen sulfide, and after that increases. Water has an anomalously high boiling point since its particles are bonded with each other by the hydrogen bonds in both its liquid as well as solid states.

Trends of their Properties

Properties of oxygen are very different from other elements of the group, but they all have 2 elec-tions in the outer s orbital, and 4 electrons in the p orbitals, usually written as s^2p^4.

The electron configurations for each element are given below,

- Oxygen: $1s^2$ $2s^2$ $2p^4$
- Sulfur: $1s^2$ $2s^2p^6$ $3s^2p^4$
- Selenium: $1s^2$ $2s^2p^6$ $3s^2p^6d^{10}$ $4s^2p^4$
- Tellurium: $1s^2$ $2s^2p^6$ $3s^2p^6d^{10}$ $4s^2p^6d^{10}$ $5s^2p^4$
- Polonium: $1s^2$ $2s^2p^6$ $3s^2p^6d^{10}$ $4s^2p^6d^{10}f^{14}$ $5s^2p^6d^{10}$ $6s^2p^4$

The trends of their properties in this group are interesting. Knowing the trend allows us to predict their reactions with other elements. Most trends are true for all groups of elements, and the group trends are due mostly to the size of the atoms and number of electrons per atom. The trends are described below:

1. The metallic properties increase in the order oxygen, sulfur, selenium, tellurium, or polonium. Polonium is essentially a metal.
2. Electronegativity, ionization energy (or ionization potential IP), and electron affinity decrease for the group as atomic weight increases.
3. The atomic radii and melting point increase.
4. Oxygen differs from sulfur in chemical properties due to its small size. The differences between O and S are more than the differences between other members.

Metallic character increases down the group, with tellurium classified as a metalloid and polonium as a metal. Melting point, boiling point, density, atomic radius, and ionic radius all increase down the group. Ionization energy decreases down the group. The most common oxidation state is -2; however, sulfur can also exist at a +4 and +6 state, and +2, +4, and +6 oxidation states are possible for Se, Te, and Po.

Table: Select properties of Group 16 elements.

	Oxygen	Sulfur	Selenium	Tellurium	Polonium
Boiling Pt (°C)	-182.962	444.674	685	989.9	962
Ionization Energy (kJ/mol)	1314	1000	941	869	812
Ionic Radius (pm)	140	184	198	221	

Oxygen

Oxygen is a gas at room temperature and 1 atm, and is colorless, odorless, and tasteless. It is the most abundant element by mass in both the Earth's crust and the human body. It is second to nitrogen as the most abundant element in the atmosphere. There are many commercial uses for oxygen gas, which is typically obtained through fractional distillation. It is used in the manufacture of iron, steel, and other chemicals. It is also used in water treatment, as an oxidizer in rocket fuel, for medicinal purposes, and in petroleum refining.

Oxygen has two allotropes, O_2 and O_3. In general, O_2 (or dioxygen) is the form referred to when talking about the elemental or molecular form because it is the most common form of the element. The O_2 bond is very strong, and oxygen can also form strong bonds with other elements. However, compounds that contain oxygen are considered to be more thermodynamically stable than O_2.

The latter allotrope, ozone, is a pale-blue poisonous gas with a strong odor. It is a very good oxidizing agent, stronger than dioxygen, and can be used as a substitute for chlorine in purifying drinking water without giving the water an odd taste. However, because of its unstable nature it disappears and leaves the water unprotected from bacteria. Ozone at very high altitudes in the atmosphere is responsible for protecting the Earth's surface from ultraviolet radiation; however, at lower altitudes it becomes a major component of smog.

Oxygen's primary oxidation states are -2, -1, 0, and -1/2 (in O_2^-), but -2 is the most common. Typically, compounds with oxygen in this oxidation state are called oxides. When oxygen reacts with metals, it forms oxides that are mostly ionic in nature. These can dissolve in water and react to form hydroxides; they are therefore called basic anhydrides or basic oxides. Nonmetal oxides, which form covalent bonds, are simple molecules with low melting and boiling points.

Compounds with oxygen in an oxidation state of -1 are referred to as peroxides. Examples of this type of compound include Na_2O_2 and BaO_2. When oxygen has an oxidation state of -1/2, as in O_2^-, the compound is called a superoxide.

Oxygen is rarely the central atom in a structure and can never bond with more than 4 elements due to its small size and its inability to create an expanded valence shell. Oxygen reacts with hydrogen to form water, which is extensively hydrogen-bonded, has a large dipole moment, and is considered an universal solvent.

There are a wide variety of oxygen-containing compounds, both organic and inorganic: oxides, peroxides and superoxides, alcohols, phenols, ethers, and carbonyl-containing compounds such as aldehydes, ketones, esters, amides, carbonates, carbamates, carboxylic acids and anhydrides.

Sulfur

Sulfur is a solid at room temperature and 1 atm pressure. It is usually yellow, tasteless, and nearly odorless. It is the sixteenth most abundant element in Earth's crust. It exists naturally in a variety of forms, including elemental sulfur, sulfides, sulfates, and organosulfur compounds. Since the 1890s, sulfur has been mined using the Frasch process, which is useful for recovering sulfur from deposits that are under water or quicksand. Sulfur produced from this process is used in a variety of ways including in vulcanizing rubber and as fungicide to protect grapes and strawberries.

Sulfur is unique in its ability to form a wide range of allotropes, more than any other element in the periodic table. The most common state is the solid S_8 ring, as this is the most thermodynamically stable form at room temperature. Sulfur exists in the gaseous form in five different forms (S, S_2, S_4, S_6, and S_8). In order for sulfur to convert between these compounds, sufficient heat must be supplied.

Two common oxides of sulfur are sulfur dioxide (SO_2) and sulfur trioxide (SO_3). Sulfur dioxide is formed when sulfur is combusted in air, producing a toxic gas with a strong odor. These two compounds are used in the production of sulfuric acid, which is used in a variety of reactions. Sulfuric acid is one of the top manufactured chemicals in the United States, and is primarily used in the manufacture of fertilizers.

Sulfur also exhibits a wide range of oxidation states, with values ranging from -2 to +6. It is often the central ion in a compound and can easily bond with up to 6 atoms. In the presence of hydrogen it forms the compound hydrogen sulfide, H_2S, a poisonous gas incapable of forming hydrogen bonds and with a very small dipole moment. Hydrogen sulfide can easily be recognized by its strong odor that is similar to that of rotten eggs, but this smell can only be detected at low, nontoxic concentrations. This reaction with hydrogen epitomizes how differently oxygen and sulfur act despite their common valence electron configuration and common nonmetallic properties.

A variety of sulfur-containing compounds exist, many of them organic. The prefix *thio-* in from of the name of an oxygen-containing compound means that the oxygen atom has been substituted with a sulfur atom. General categories of sulfur-containing compounds include thiols (mercaptans), thiophenols, organic sulfides (thioethers), disulfides, thiocarbonyls, thioesters, sulfoxides, sulfonyls, sulfamides, sulfonic acids, sulfonates, and sulfates.

Selenium

Selenium appears as a red or black amorphous solid, or a red or grey crystalline structure; the latter is the most stable. Selenium has properties very similar to those of sulfur; however, it is more metallic though it is still classified as a nonmetal. It acts as a semiconductor and therefore is often used in the manufacture of rectifiers, which are devices that convert alternating currents to direct currents. Selenium is also photoconductive, which means that in the presence of light the electrical conductivity of selenium increases. It is also used in the drums of laser printers and copiers. In addition, it has found increased use now that lead has been removed from plumbing brasses.

It is rare to find selenium in its elemental form in nature; it must typically be removed through a refining process, usually involving copper. It is often found in soils and in plant tissues that have bioaccumulated the element. In large doses, the element is toxic; however, many animals require it as an essential micronutrient. Selenium atoms are found in the enzyme glutathione peroxidase, which destroys lipid-damaging peroxides. In the human body it is an essential cofactor in maintaining the function of the thyroid gland. In addition, some research has shown a correlation between selenium-deficient soils and an increased risk of contracting the HIV/AIDS virus.

Tellurium

Tellurium is the metalloid of the oxygen family, with a silvery white color and a metallic luster similar to that of tin at room temperature. Like selenium, it is also displays photoconductivity. Tellurium is an extremely rare element, and is most commonly found as a telluride of gold. It is often used in metallurgy in combination with copper, lead, and iron. In addition, it is used in solar panels and memory chips for computers. It is not toxic or carcinogenic; however, when humans are exposed to too much of it they develop a garlic-like smell on their breaths.

Polonium

Polonium is a very rare, radioactive metal. There are 33 different isotopes of the element and all of the isotopes are radioactive. It exists in a variety of states, and has two metallic allotropes. It dissolves easily into dilute acids. Polonium does not exist in nature in compounds, but it can form synthetic compounds in the laboratory. It is used as an alloy with beryllium to act as a neutron source for nuclear weapons.

Polonium is a highly toxic element. The radiation it emits makes it very dangerous to handle. It can be immediately lethal when applied at the correct dosage, or cause cancer if chronic exposure to the radiation occurs. Methods to treat humans who have been contaminated with polonium are still being researched, and it has been shown that chelation agents could possibly be used to decontaminate humans.

The Group 17 Elements

The group 17 elements include fluorine(F), chlorine(Cl), bromine(Br), iodine(I) and astatine(At) from the top to the bottom. They are called "halogens" because they give salts when they react with metals. So, now you know what halogens are. Let's focus on the electronic configuration of these elements.

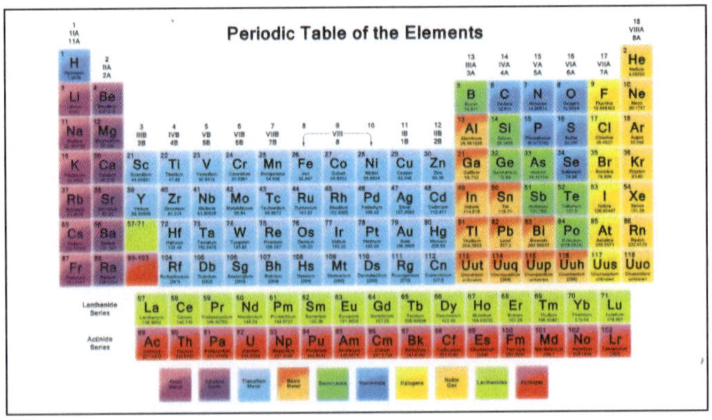

Electronic Configuration of Group 17 Elements

The valence shell electronic configuration of these electrons is ns^2np^5. Thus, there are 7 electrons in the outermost shell of these elements. The element misses out on the octet configuration by one electron. Thus, these elements look out to either lose one electron and form a covalent bond or gain one electron and form an ionic bond. Therefore, these are very reactive non-metals.

Atomic Properties

Let us focus at the various atomic properties of the group 17 elements. We will speak about the ionic and atomic radii, ionization enthalpy and more.

Ionic and Atomic Radii

The nuclear and atomic radii of these elements keep on increasing as we move down the group. This happens because of the addition of an extra energy level. They have the minimal atomic radii compared to the other elements in the related periods. This can be attributed to the fact that their atomic charge is quite powerful.

Ionisation Enthalpy

These elements have higher ionization enthalpy. This value keeps on diminishing as we move down the group. This happens because of the increase in the size of the nucleus. However, it is interesting to note that fluorine has the highest ionization enthalpy than any other halogen, thanks to its minute size.

Electron Gain Enthalpy

The electron gain enthalpy of these elements becomes less negative upon moving down the group. Fluorine has lesser enthalpy than chlorine. We can attribute it to the small size and the smaller 2p sub-shell of the atom of fluorine.

Electro-negativity

The halogens exhibit high electro-negativity values. However, it diminishes slowly on moving down the group from fluorine to iodine. This can be attributed to the increase in nuclear radii upon moving down the group.

Physical Properties

Let us focus at the various physical properties of these halogens.

- Physical state: The group 17 elements are found in diverse physical states. For example, Fluorine and Chlorine are gases. On the other hand, Bromine is a liquid and Iodine is solid.
- Color: These elements have a variety of Colors. For example, while Fluorine is pale yellow in Color, Iodine is dark violet in Color.
- Solubility: Florine and Chlorine are soluble in water. On the other hand, Bromine and Iodine are very less soluble in water.
- Melting and boiling points: Melting and boiling points of these elements increase as we move down the group from Fluorine to Iodine. Thus, fluorine has the lowest boiling and melting points.

Chemical Properties

Let us focus at some of the chemical properties of these elements.

Oxidising Power

All the halogens are great oxidizing agents. Of the list, fluorine is the most powerful oxidizing agent. It is capable of oxidizing all the halide particles to halogen. The oxidizing power reduces as we move down the group. The halide particles also act as reducing agents. However, their reducing capacity decreases down the group as well.

Reaction with Hydrogen

All halogens react with hydrogen and produce acidic hydrogen halides. The acidity of these hydrogen halides reduces from HF to HI. Fluorine reacts violently and chlorine requires the sunlight. On the other hand, bromine reacts upon heating and iodine needs a catalyst.

Reaction with Oxygen

Halogens react with oxygen to form oxides. However, it has been found that the oxides are not steady. Beside oxides, halogens also form a number of halogen oxoacids and oxoanions.

Reaction with Metals

As halogens are very reactive, they react with most of the metals instantly and form the resulting metal halides. For example, sodium reacts with chlorine gas and forms sodium chloride. This process is an exothermic one and gives out a bright yellow light and a lot of heat energy.

$$2Na_{(s)} + Cl_{2(g)} \rightarrow 2NaCl_{(s)}$$

Metal halides are ionic in nature. This is because of the high electronegative nature of the halogens and high electropositivity of the metals. This ionic character of the halides reduces from fluorine to iodine.

The Group 18 Elements

The noble gases (Group 18) are located in the far right of the periodic table and were previously referred to as the "inert gases" due to the fact that their filled valence shells (octets) make them extremely nonreactive. The noble gases were characterized relatively late compared to other element groups.

The first person to discover the noble gases was Henry Cavendish in the late 180th century. Cavendish distinguished these elements by chemically removing all oxygen and nitrogen from a container of air. The nitrogen was oxidized to NO_2 by electric discharges and absorbed by a sodium hydroxide solution. The remaining oxygen was then removed from the mixture with an absorber. The experiment revealed that 1/120 of the gas volume remained un-reacted in the receptacle. The second person to isolate, but not typify, them was William Francis. Francis noted the formation of gas while dissolving uranium minerals in acid.

Argon

In 1894, John William Strutt discovered that chemically-obtained pure nitrogen was less dense than the nitrogen isolated from air samples. From this breakthrough, he concluded that another, unknown gas was present in the air. With the aid of William Ramsay, Strutt managed to replicate and modify Cavendish's experiment to better understand the inert component of air in his original experiment. The researchers' procedure differed from the Cavendish procedure: they removed the oxygen by reacting it with copper, and removed the nitrogen in a reaction with magnesium. The remaining gas was properly characterized and the new element was named "argon," which originates from the Greek word for "inert."

Helium

Helium was first discovered in 1868, manifesting itself in the solar spectrum as a bright yellow line with a wavelength of 587.49 nanometers. This discovery was made by Pierre Jansen. Jansen initially assumed it was a sodium line. However, later studies by Sir William Ramsay (who isolated helium on Earth by treating a variety of rare elements with acids) confirmed that the bright yellow line from his experiment matched up with that in the spectrum of the sun. From this, British physicist William Crookes identified the element as helium.

Neon, Krypton and Xenon

These three noble gases were discovered by Morris W. Travers and Sir William Ramsay in 1898. Ramsay discovered neon by chilling a sample of the air to a liquid phase, warming the liquid, and capturing the gases as they boiled off. Krypton and xenon were also discovered through this process.

Radon

In 1900, while studying the decay chain of radium, Friedrich Earns Dorn discovered the last gas in Group 18: Radon. In his experiments, Dorn noticed that radium compounds emanated radioactive gas. This gas was originally named niton after the Latin word for shining, "nitens". In 1923, the International Committee for Chemical Elements and International Union of Pure Applied Chemistry (IUPAC) decided to name the element radon. All isotopes of radon are radioactive. Radon-222 has the longest half-life at less than 4 days, and is an alpha-decay product of Radium-226 (part of the U-238 to Pb-206 radioactive decay chain).

The Electron Configurations for Noble Gases

- Helium $1s^2$
- Neon [He] $2s^2$ $2p^6$
- Argon [Ne] $3s^2$ $3p^6$
- Krypton [Ar] $3d^{10}$ $4s^2$ $4p^6$
- Xenon [Kr] $4d^{10}$ $5s^2$ $5p^6$
- Radon [Xe] $4f^{14}$ $5d^{10}$ $6s^2$ $6p^6$

Table: Trends within Group 18

	Atomic #	Atomic mass	Boiling point (K)	Melting point (K)	1st Ionization (E/kJ mol $^{-1}$)	Density (g/dm^3)	Atomic radius (pm)
He	2	4.003	4.216	0.95	2372.3	0.1786	31
Ne	10	20.18	27.1	24.7	2080.6	0.9002	38
Ar	18	39.948	87.29	83.6	1520.4	1.7818	71
Kr	36	83.3	120.85	115.8	1350.7	3.708	88
Xe	54	131.29	166.1	161.7	1170.4	5.851	108
Rn	86	222.1	211.5	202.2	1037.1	9.97	120

Atomic and Physical Properties

- Atomic mass, boiling point, and atomic radii increase down a group in the periodic table.
- The first ionization energy decreases down a group in the periodic table.
- The noble gases have the largest ionization energies, reflecting their chemical inertness.
- Down Group 18, atomic radius and interatomic forces increase resulting in an increased melting point, boiling point, enthalpy of vaporization, and solubility.
- The increase in density down the group is correlated with the increase in atomic mass.
- Because the atoms increase in atomic size down the group, the electron clouds of these non-polar atoms become increasingly polarized which leads to weak van der waals forces

among the atoms. Thus, the formation of liquids and solids is more easily attainable for these heavier elements because of their melting and boiling points.

- Because noble gases' outer shells are full, they are extremely stable, tending not to form chemical bonds and having a small tendency to gain or lose electrons.
- Under standard conditions all members of the noble gas group behave similarly.
- All are monotomic gases under standard conditions.
- Noble gas atoms, like the atoms in other groups, increase steadily in atomic radius from one period to the next due to the increasing number of electrons.
- The size of the atom is positively correlated to several properties of noble gases. The ionization potential decreases with an increasing radius, because the valence electrons in the larger noble gases are further away from the nucleus; they are therefore held less tightly by the atom.
- The attractive force increases with the size of the atom as a result of an increase in polarizability and thus a decrease in ionization potential.
- Overall, noble gases have weak interatomic forces, and therefore very low boiling and melting points compared with elements of other groups.

For covalently-bonded diatomic and polyatomic gases, heat capacity arises from possible translational, rotational, and vibrational motions. Because monatomic gases have no bonds, they cannot absorb heat as bond vibrations. Because the center of mass of monatomic gases is at the nucleus of the atom, and the mass of the electrons is negligible compared to the nucleus, the kinetic energy due to rotation is negligible compared to the kinetic energy of translation (unlike in di- or polyatomic molecules where rotation of nuclei around the center of mass of the molecule contributes significantly to the heat capacity). Therefore, the internal energy per mole of a monatomic noble gas equals its translational contribution, $\frac{3}{2}RT$, where R is the universal gas constant and T is the absolute temperature.

For monatomic gases at a given temperature, the average kinetic energy due to translation is practically equal regardless of the element. Therefore at a given temperature, the heavier the atom, the more slowly its gaseous atoms move. The mean velocity of a monatomic gas decreases with increasing molecular mass, and given the simplified heat capacity situation, noble gaseous thermal conductivity decreases with increasing molecular mass.

Applications of Noble Gases

Helium

Helium is used as a component of breathing gases due to its low solubility in fluids or lipids. This is important because other gases are absorbed by the blood and body tissues when under pressure during scuba diving. Because of its reduced solubility, little helium is taken into cell membranes; when it replaces part of the breathing mixture, helium causes a decrease in the narcotic effect of the gas at far depths. The reduced amount of dissolved gas in the body means fewer gas bubbles form, decreasing the pressure of the ascent. Helium and Argon are used to shield welding arcs and the surrounding base metal from the atmosphere.

Helium is used in very low temperature cryogenics, particularly for maintaining superconductors (useful for creating strong magnetic fields) at very low temperatures. Helium is also the most common carrier gas in gas chromatography.

Neon

Neon has many common and familiar applications: neon lights, fog lights, TV cine-scopes, lasers, voltage detectors, luminous warnings, and advertising signs. The most popular application of neon is the neon tubing used in advertising and elaborate decorations. These tubes are filled with neon and helium or argon under low pressure and submitted to electrical discharges. The color of emitted light is depends on the composition of the gaseous mixture and on the color of the glass of the tube. Pure Neon within a colorless tube absorbs red light and reflects blue light. This reflected light is known as fluorescent light.

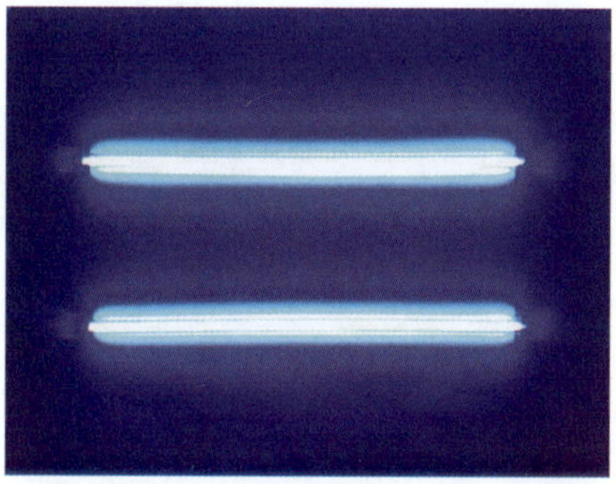

One of the many colors of neon lights.

Argon

Argon has a large number of applications in electronics, lighting, glass, and metal fabrications. Argon is used in electronics to provide a protective heat transfer medium for ultra-pure silicon crystal semiconductors and for growing germanium. Argon can also fill fluorescent and incandescent light bulbs, creating the blue light found in "neon lamps." By utilizing argon's low thermal conductivity, window manufacturers provide a gas barrier needed to produce double-pane insulated windows. This insulation barrier improves the windows' energy efficiencies. Argon also creates an inert gas shield during welding, flushes out melted metals to eliminate porosity in casting, and provides an oxygen- and nitrogen-free environment for annealing and rolling metals and alloys.

Argon plasma light bulb.

Krypton

Similarly to argon, krypton can be found in energy efficient windows. Because of its superior thermal efficiency, krypton is sometimes chosen over argon for insulation. It is estimated that 30% of energy efficient windows sold in Germany and England are filled with krypton; approximately 1.8 liters of krypton are used in these countries. Krypton is also found in fuel sources, lasers and headlights. In lasers, krypton functions as a control for a desired optic wavelength. It is usually mixed with a halogen (most likely fluorine) to produce excimer lasers. Halogen sealed beam headlights containing krypton produce up to double the light output of standard headlights. In addition, Krypton is used for high performance light bulbs, which have higher color temperatures and efficiency because the krypton reduces the rate of evaporation of the filament.

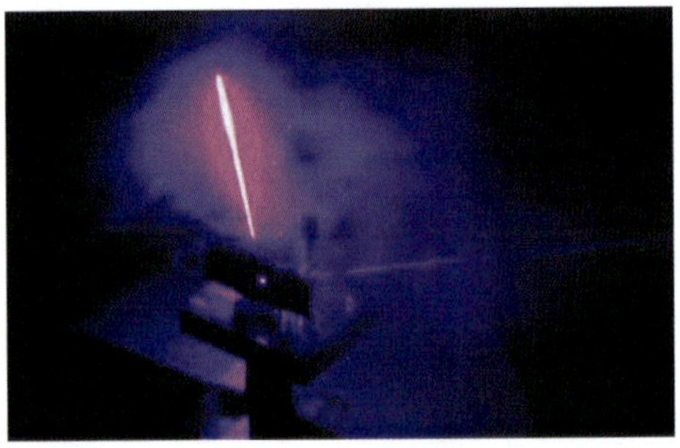

Krypton laser.

Xenon

Xenon has various applications in incandescent lighting, x-ray development, plasma display panels (PDPs), and more. Incandescent lighting uses xenon because less energy can be used to obtain the same light output as a normal incandescent lamp. Xenon has also made it possible to obtain better x-rays with reduced amounts of radiation. When mixed with oxygen, it can enhance the contrast in CT imaging. These applications have had great impact on the health care industries. Plasma display panels (PDPs) using xenon as one of the fill gases may one day replace the large picture tubes in television and computer screens.

Nuclear fission products may include several radioactive isotopes of xenon, which absorb neutrons in nuclear reactor cores. The formation and elimination of radioactive xenon decay products are factors in nuclear reactor control.

Radon

Radon is reported as the second most frequent cause of lung cancer, after cigarette smoking. However, it also has beneficial applications in radiotherapy, arthritis treatment, and bathing. In radiotherapy, radon has been used in implantable seeds, made of glass or gold, primarily used to treat cancers. It has been said that exposure to radon mitigates auto-immune diseases such as arthritis. Some arthritis sufferers have sought limited exposure to radioactive mine water and radon to relieve their pain. "Radon Spas" such as Bad Gastern in Austria and Onsen in Japan offer a therapy in which people sit for minutes to hours in a high-radon atmosphere, believing that low doses of radiation will boost up their energy.

The D Block Elements

D block elements are the elements which can be found from the third group to the twelfth group of the modern periodic table. The valence electrons of these elements fall under the d orbital. D block elements are also referred to as transition elements or transition metals. The first three rows of the d block elements which correspond to the 3d, 4d, and 5d orbitals.

Elements having electrons (1 to 10) present in the d-orbital of the penultimate energy level and in the outer most 's' orbital (1-2) are d block elements. Although electrons do not fill up 'd' orbital in the group 12 metals, their chemistry is similar in many ways to that of the preceding groups, and so considered as d block elements.

These elements typically display metallic qualities such as malleability and ductility, high values of electrical conductivity and thermal conductivity, and good tensile strength. There are four series in the d block corresponding to the filling up of 3d, 4d, 5d or 6d orbitals.

- 3d- Sc, Ti, V, Cr, Mn, Fe, Co, Ni, Cu, Zn
- 4d- Y, Zr, Nb, Mo, Tc, Ru, Rh, Pd, Ag, Cd
- 5d- La, Hf, Ta, W, Re, Os, Ir, Pt, Au, Hg
- 6d- incomplete.

There are 10 elements filling up the 'd' orbital in each series.

Position of D Block Elements in Periodic Table									
21 Sc	22 Ti	23 V	24 Cr	25 Mn	26 Fe	27 Co	28 Ni	29 Cu	30 Zn
39 Y	40 Zr	41 Nb	42 Mo	43 Tc	44 Ru	45 Rh	46 Pd	47 Ag	48 Cd
57 La	72 Hf	73 Ta	74 W	75 Re	76 Os	77 Ir	78 Pt	79 Au	80 Hg

D block elements, occupy columns 3 to 12 and may have atoms of elements with completely filled 'd' orbital. IUPAC defines a transition metal as "an element whose atom or its cations has a partially filled d sub-shell.

D Block Elements also Called Transition Elements

Transition elements occupy groups 4–11. Scandium and yttrium of group 3, having a partially filled d subshell in the metallic state also are considered as transitions elements. Elements like Zn, CD and Hg of the 12 column of the d block have completely filled d-orbital and hence are not considered as transition elements.

Transition Elements are so named, indicating their positioning and transition of properties between, s and p block elements. So, all the transition metals are d block elements but all d block elements are not transition elements.

Properties of Transition Metals

- Electrons added to the 'd' sub-orbitals that lie between their (n+1) s and (n+1) p sub-orbitals.
- Placed between s and p block elements in the periodic table.
- Properties between s and p-block elements.

Electronic Configuration of D Block Elements

D block Elements have a general electronic configuration of $(n\text{-}1)d^{1\text{-}10} ns^{1\text{-}2}$. These elements can find stability in half-filled orbitals and completely filled d orbitals. An example of this would be the electronic configuration of chromium, which has half-filled d and s orbitals in its configuration – 3d54s1. The electronic configuration of copper is another such example. Copper displays an electronic configuration of 3d104s1 and not 3d94s2.

This can be attributed to the relative stability of the completely filled d orbital. Zinc, Mercury, Cadmium, and Copernicium exhibit completely filled orbitals in their ground states and in their general oxidation states as well. For this reason, these metals are not considered as transition elements whereas the others are d block elements.

- The electronic configuration for period 4, transition elements is (Ar) $4s^{1\text{-}2}\, 3d^{1\text{-}10}$.
- The electronic configuration for period 5, transition elements is (Kr) $5s^{1\text{-}2}\, 4d^{1\text{-}10}$.
- The electronic configuration for period 6, transition elements is (Xe) $4s^{1\text{-}2}\, 3d^{1\text{-}10}$.

Along the period, from left to right, electrons are added to the 3d subshell as per the Aufbau principle and Hund's rule of multiplicity.

1st transition series	Sc	Ti	V	Cr	Mn	Fe	Co	Ni	Cu	Zn
	$4s^2 3d^1$	$4s^2 3d^2$	$4s^2 3d^3$	$4s^1 3d^5$	$4s^2 3d^5$	$4s^2 3d^6$	$4s^2 3d^7$	$4s^2 3d^8$	$4s^1 3d^{10}$	$4s^2 3d^{10}$
2nd transition series	Y	Zr	Nb	Mo	Tc	Ru	Rh	Pd	Ag	Cd
	$5s^2 4d^1$	$5s^2 4d^2$	$5s^1 4d^4$	$5s^1 4d^5$	$5s^2 4d^5$	$5s^1 4d^7$	$5s^1 4d^8$	$5s^0 4d^{10}$	$5s^1 4d^{10}$	$5s^2 4d^{10}$
3rd transition series	La	Hf	Ta	W	Re	Os	Ir	Pt	Au	Hg
	$6s^2 5d^1$	$6s^2 5d^2$	$6s^2 5d^3$	$6s^2 5d^4$	$6s^2 5d^5$	$6s^2 5d^6$	$6s^2 5d^7$	$6s^1 5d^9$	$6s^1 5d^{10}$	$6s^2 5d^{10}$

Anomalies do occur in all the series, which can be explained from the following considerations:

- The energy gap between the ns and (n-1) d orbitals.
- Pairing energy for the electrons in s-orbital.
- Stability of half-filled orbitals to the partly filled orbitals.

Chromium has $4s^1 3d^5$ electron configuration rather than the $4s^2 3d^4$ configuration and copper $4s^1 3d^{10}$ rather than $4s^2 3d^9$. These anomalies in the first transition series can be understood from the stability of half-filled orbitals compared to the partly filled orbitals.

In the second series transition metals, from niobium, electron presence in d orbitals appears to be preferred than being shared in s orbitals. Between the available s and d orbitals, the electron can either go for sharing in s-orbital or excited to d-orbital. Obviously, the choice depends on the repulsive energy it has overcome on sharing and the energy gap between the s and d-orbitals.

In the second series, s and d-orbital have the almost same energy, because of which electrons prefer to occupy the d-orbital. So from niobium, s-orbital has mostly only one electron. Third series transition metals, on the other hand, has more paired s configuration even at the expense of half-filled orbitals (Tungsten- $6s^2 5d^4$). This series comes after filling up of 4f orbitals and resulting lanthanide contraction.

The reduced size results in high shielding of d orbitals by the 'f' electron. This shielding increases the energy gap between the s and 5d orbitals such that pairing energy is less than the excitation. Excitation of the electron does not take place in tungsten, in spite of the stability possible because of half-filled orbitals.

Atomic and Ionic Radii of D Block Elements

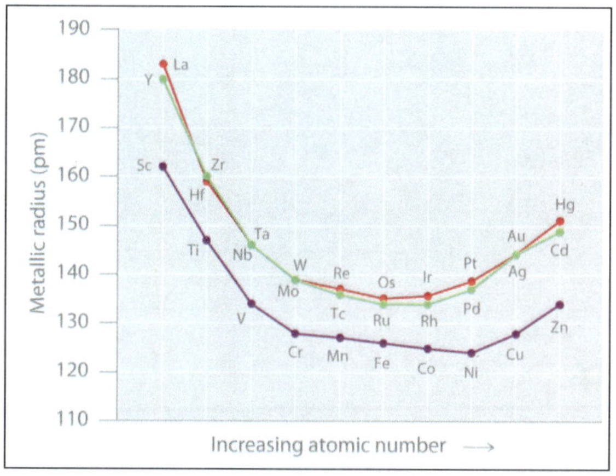

Metallic Radii of 1st, 2nd, and 3rd Row Transition Metals.

Atomic and ionic radii of elements of all three-transition series:

- Decreases rapidly, from column 3 to 6;
- Remains steady, from column 7 to 10; and
- Start increasing from column 11 to 12.

For example, in the first transition series, atomic radii, the decrease is more from Sc to Cr (group 3 to 6), is almost same for Mn, Fe, Co, Ni (group 7, 8 9 & 10) and increase in cu and Zn.

- The larger decrease in atomic radii, in column 3 to 6 elements is due to the increase in effective nuclear charge but poor shielding because of the smaller number of d-electrons.
- In elements of column 7 to 10 increasing effective nuclear charge is balanced by the repulsion between the shared d electrons so that radii remain the same.
- In the case of 11 and 12 columns elements, the d orbital is full with ten electrons and shield the electrons present in the higher s-orbital. So, group 11 & 12 elements like Cu & Zn have bigger size than its earlier elements in the block.

Since electrons occupy a higher orbital, radii of the third series are to be more than the second series elements. But radii of both series are almost the same. In the third series elements, 5d orbitals are filled only after the filling up of 4f orbitals, which increases the effective nuclear charge by 14 units.

This higher nuclear charge leads to the larger shrinkage of radii known as Lanthanide contraction. Increase in radii due to the higher orbital will be effectively neutralized by the increase the nuclear effective charge. So, radii of second and third series elements have same atomic radii. For example, Niobium and hafnium have almost the same atomic radii.

Properties of D Block Elements

Ionization Energy of D Block Elements

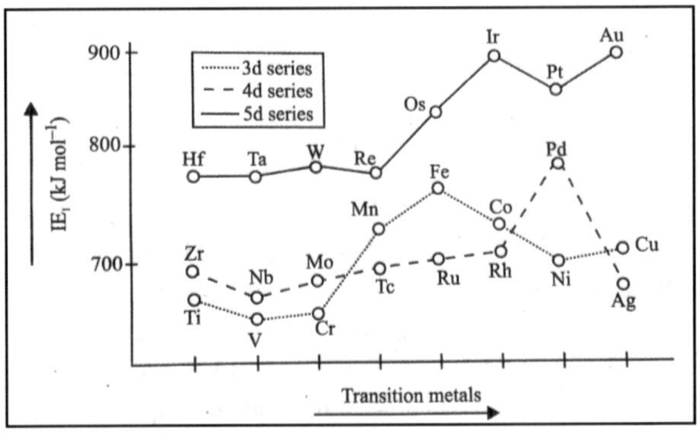

Ionization enthalpies of transition metals.
IE_1 of 5d metals are higher than those of 3d and 4d metals.

Ionization energy is the energy needed to remove the valence electron from the atom/ion and is directly related to the force of attraction on the electron. Hence larger the nuclear charge and smaller the radii of the electron larger will be the ionization energy (IE). Ionization Energy also will be more for half-filled and fully filled orbitals.

Ionization Energy of the d block elements is larger than s-block and smaller than the p-block elements, between which, they are placed. In the first series, except chromium and copper first Ionization Energy involves removal from filled s-orbital. Among them, Ionization Energy of d block elements increases with the increase in atomic number up to Fe.

In Co and Ni, increasing sharing of d-electrons compensate for the atomic number increase resulting in the decrease of Ionization Energy. Copper and zinc show increasing IE, as s -block elements. In the second series, elements from Niobium have single electrons in the s-orbital.

Hence, they show a gradual increase in IE with increasing atomic number. Palladium, on the other hand, has completed d-shell and no electron in s-shell. So, Pd shows the maximum IE. Because of lanthanide contraction, the attraction of electrons by the nuclear charge is much higher and hence IE of 5d elements are much larger than 4d and 3d. In 5d series, all elements except Pt and Au have filled s-shell.

Elements from Hafnium to rhenium have same IE and after IE increases with the number of shared d-electrons such that Iridium and Gold have the maximum IE.

Metallic Character

D block elements show typical metallic behavior of high tensile strength, malleability, ductility, electrical and thermal conductivity, metallic lustre and crystallize in bcc/ccp/hcp structures.

They are very hard and have a high enthalpy of atomization and low volatility except for Copper. Hardness increases with the number of unpaired electrons. Hence Cr, Mo and W are very hard metals among d block elements. The group-12 elements (Zn, Cd and Hg) show the exception in this regard also.

Oxidation States of D Block Elements

Oxidation state is a hypothetical state, where the atom appears to release or gain electrons more than the usual valency state. It is still useful in explaining the properties of the atom/ion. Transition elements/ions may have electrons in both s and d-orbitals.

Since the energy difference between s and d-orbital are small, both the electrons can involve in ionic and covalent bond formation and hence exhibit multiple(variable) valency states (oxidation states).

Each transition element can hence exhibit a minimum oxidation state corresponding to the number of s-electrons and maximum oxidation state equivalent to the total number of electrons available in both s and d-orbitals. In between oxidation states also become possible.

Sc	+2,+3	+3	+3	Y	+2,+3	+3	+3	La	+2,+3	+3	+3	
Ti	+2,+3,+4,	+2	+4	Zr	+2,+3,+4,	+2	+4	Hf	+2,+3,+4,	+4	+4	
V	+2,+3,+4,+5,	+2	+5	Nb	+2,+3,+4,+5,	+2	+5	Ta	+2,+3,+4,+5,	+4	+5	
Cr	+2,+3,+4,+5,+6,	+1	+2	+6	Mo	+2,+3,+4, +5,+6,	+4	+6	W	+2,+3,+4, +5,+6,	+4	+6
Mn	+2,+3,+4,+5, +6,+7	+2	+7	Tc	+2,+3,+4,+5, +6,+7	+4	+7	Re	+2,+3,+4,+5, +6,+7	+4	+7	
Fe	+2,+3,+4,+5,+6,	+2	+6	Ru	+2,+3,+4,+5,+6, +7,+8	+4	+8	Os	+2,+3,+4,+5, +6,+7,+8	+4	+8	
Co	+2,+3,+4,	+2	+4	Rh	+2,+3,+4,	+3	+4	Ir	+2,+3,+4,	+4	+4	
Ni	+2,+3,+4,	+2	+4	Pd	+2,+3,+4,	+2	+4	Pt	+2,+3,+4,	+4	+4	
Cu	+1,+2,	+1	+2	+2	Ag	+1,+2,	+1	+2	Au	+1,+2,	+1	+2
Zn	+2,	+2	+2	Cd	+2,	+2	+2	Hg	+2,	+1	+2	+2

Trends in the Oxidation States

1. The minimum Oxidation state of 1 is shown by Cr, Cu, Ag, Au and Hg.

2. More stable Oxidation state increases in the order 3d < 4d < 5d. 3d series elements are most stable in +2; 4d series in +2 and +4 and 5d series in +4. Cr^{6+} and Mn^{7+} (of 3d) are not stable in their higher OS. Compounds containing them, CrO_4^{2-} and MnO_4^- are very reactive and strong oxidizing agents.

While Mo^{6+} and Tc^{7+} (of 4d) are stable in their higher OS. Compounds containing them, MO_4^{2-} and TcO_4^- are unreactive and stable. Similarly, W^{6+} and Re^{7+} (of 5d) are stable in their higher OS. Compounds containing them, WO_4^{2-} and ReO_4^- are unreactive and stable.

Cations of the second and third-row transition metals in lower oxidation states (+2 and +3) are much more easily oxidized than the corresponding ions of the first-row transition metals. For example, the most stable compounds of chromium are those of Cr(III), but the corresponding Mo(III) and W(III) compounds are highly reactive.

In fact, they are often pyrophoric, bursting into flames on contact with atmospheric oxygen. the heavier elements in each group form stable compounds in higher oxidation states that have no analogues with the lightest member of the group.

3. Strongly oxidizing, high oxidation number elements form compounds of oxides and fluorides and not bromides and iodides.

Vanadium form only VO_4^-, CrO_4^{2-}, MnO_4^-, VF_5, VCl_5, VBr_3, VI_3 and not VBr_5, VI_5, V^{5+} oxidizes Br^- and I^- to Br_2 and I_2 but not fluoride because of its high electronegativity and small size.

Similarly, strongly reducing, low oxidation number elements form bromides and iodides and not oxides and fluorides.

4. Maximum oxidation state equal to the s and d-electrons is exhibited by middle-order elements in each series. Thus, manganese in 3d series has +7, Ru in 4d and Os in 5d possess +8 maximum oxidation state.

5. Elements may show all the Oxidation states in between the minimum and maximum.

6. Elements in their lower oxidation states will be ionic and basic (TiO, VO, CrO, MnO, $TiCl_2$ and VCl_2) in-between state amphoteric (Ti_2O_3, V_2O_3, Mn_2O_3, CrO_3, Cr_2O_3, $TiCl_3$, VCl_3) and higher oxidation state covalent and acidic (V_2O_5, MnO_3, Mn_2O_7, VCl_4 and $VOCl_3$).

7. Lower oxidation state may get stabilized by back bonding in complexes. $Ni(CO)_4$, $Fe(CO)_5$, $[Ag(CN)_2]^-$, $[Ag(NH_3)_2]^+$

Lower oxidation states in these metals are stabilised by ligands like CO, which are pi-electron donors, whereas the higher oxidation states are stabilised by electronegative elements like Fluorine(F) and Oxygen(O). Hence the high oxidation compounds of these metals are mainly fluorides and oxides.

8. Relative stabilities of the oxidation states depend on many factors, like, the stability of the resulting orbital, IE, electronegativity, enthalpy of atomization, enthalpy of hydration, etc.

- Ti^{4+} ($3d^0$) is more stable than Ti^{3+}($3d^1$). Mn^{2+} ($3d^5$) is more stable than Mn^{3+}($3d^4$).
- Ionization energies contribute to the relative stability of transition metal compounds (ions). For example, Ni^{2+} compounds are thermodynamically more stable than Pt^{2+}, Whereas, Pt^{4+} compounds are more stable than Ni^{4+}. The relative stabilities can be explained as follows,

Metal	(IE1+IE2) $kJmol^{-1}$,	(IE3+IE4) $kJmol^{-1}$,	Etotal, =(=IE1+IE2 +IE3+IE4) kJ mol^{-1}
Ni	2490	8800	11290
Pt	2660	6700	9360

Thus, the ionization of Ni to Ni^{2+} requires lesser energy (2490 kJ mol^{-1}) as compared to the energy required for the production of Pt^{2+} (2660 kJ mol^{-1}). Therefore, Ni^{2+} compounds are thermodynamically more stable than Pt^{2+} compounds.

On the other hand, the formation of Pt^{4+} requires lesser energy (9360 kJ mol^{-1}) as compared to that required for the formation of Ni^{4+} (11290 kJ/mol). Therefore, Pt^{4+} compounds are more stable than Ni^{4+} compounds. This is supported by the fact that $[PtCl6]^{2+}$ complex ion is known, while the corresponding ion for nickel is not known.

9. In p-block, the heavier elements prefer lower oxidation states due to what is called the inert pair effect. But in the case of d block elements, the higher oxidation states are more stable for heavier members in a group.

Electrode Potential in D Block Elements

Relative stabilities of transition metal ions in different oxidation states in the aqueous medium can be predicted from the electrode potential data. The oxidation state of a cation for which ΔH(ΔHsub + lE + ΔHhyd) or E° is more negative (for less positive) will be more stable.

- E° becomes less negative along the series indicating higher stability of the reduced state.
- Compared to first and second group metals, transition elements have low E°.

Physical Properties of D Block Elements

Density: Among the transition series, the trend in density will be reverse of atomic radii, i.e. density increase remains almost the same and then decreases along the period.

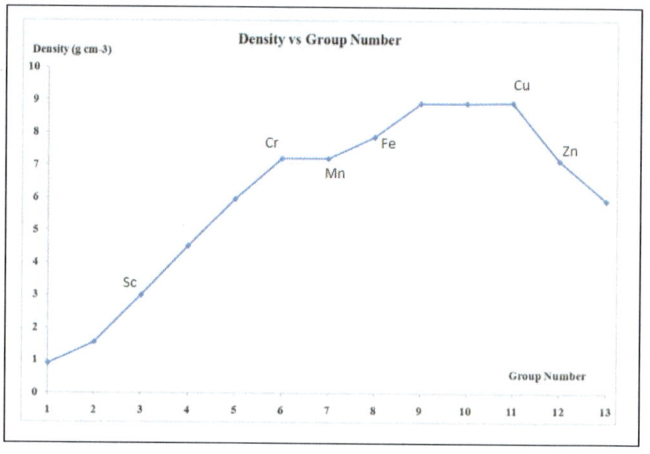

Down the column density of 4d series is larger than 3d. Because of lanthanide contraction and a larger decrease in atomic radii and hence the volume density of 5d series transition elements are double than 4d series.

In the 3d series, scandium has the lowest density and copper highest density. Osmium (d=22.57g cm^{-3}) and Iridium (d=22.61g cm^{-3}) of 5d series have the highest density among all d block elements.

Some relative radii of d block elements are Fe < Ni < Cu, Fe < Cu < Au, Fe < Hg < Au.

D Block Elements have High Melting and Boiling Point

Unpaired electrons and the empty or partially filled d-orbitals form covalent bonding in addition to the metallic bonding by s-electrons. Because of such strong bonding, d-block elements have high melting and boiling points than s and p block elements. This trend goes till d^5 configuration and then decreases as more electrons get paired in the d-orbital.

- Cr, Mo and W possess the highest melting at boiling point in their series of elements.
- Manganese (Mn) and Technetium (Tc) have half-filled configuration resulting in weak metallic bonding and abnormally low melting and boiling points.
- Group 12, Zn, Cd and Hg have no unpaired d-electrons and hence no covalent bonding. Their melting and boiling point will be the lowest in their series.

Mercury – the liquid metal: Mercury is the only metal that exists in its liquid state at room temperature. 6s valence electrons of Mercury are more closely pulled by the nucleus (lanthanide contraction) such that outer s-electrons are less involved in metallic bonding.

What Transition Elements are Considered Noble Metals?

The Three Transition Series

- The ionization energies of elements increase very slowly across a given row.
- From left of 3d series to the right corner 5d transition elements, density, electronegativity, electrical and thermal conductivities increase, while enthalpies of hydration of the metal cations decrease in magnitude.

This indicates that the transition metals become steadily less reactive and more "noble" in character. The relatively high ionization energies, increasing electronegativity, and decreasing low enthalpies of hydration make metals (Pt, Au) in the lower right corner of the d block as unreactive that they are often called the "noble metals."

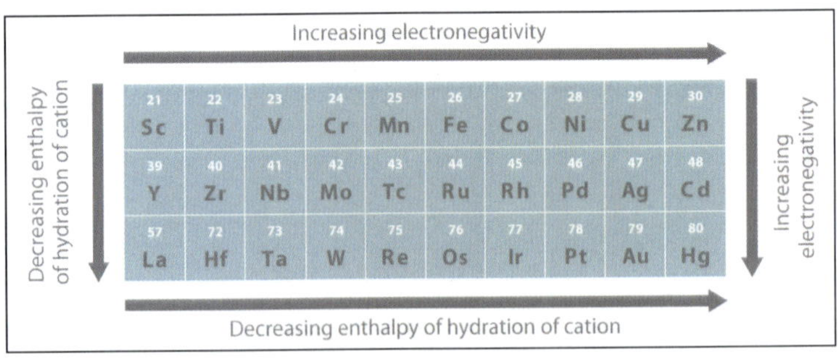

Magnetic Properties of D Block Elements

Materials are classified by their interaction with the magnetic field as:

- Diamagnetic: If repelled;
- Paramagnetic: If attracted;

- Ferromagnetic: If it can retain the larger magnetic nature even in the absence of magnetic field;
- Paired electrons cause diamagnetism: Unpaired electrons result in para-magnetism and aligned together unpaired electrons produce ferromagnetism. D block elements and their ions exhibit this behavior depending on the unpaired electrons;
- Unpaired electrons contribute to 'orbital magnetic moment' and 'spin magnetic moment'. However, for 3d series, the orbital angular moment is negligible and the approximate spin-only magnetic moment is given by the formula:

  $\mu = \sqrt{[4s (s + 1)]} = \sqrt{[n (n + 1)]}$ BM.

Where 'S' is the total spin and 'n' is the number of unpaired electrons. Its unit is Bohr Magneton (BM). For higher d-series, the actual magnetic moment includes components from the orbital moment in addition to the spin moment. Chromium and molybdenum possess maximum number (6) of unpaired electrons and magnetic moment.

Ion	Outer configuration	No. of unpaired electrons	Magnetic moment (BM)	
Calculated	observed			
Sc^{3+}	$3d^0$	0	0	0
Ti^{3+}	$3d^1$	1	1.73	1.75
Ti^{2+}	$3d^2$	2	2.84	2.86
V^{2+}	$3d^3$	3	3.87	3.86
Cr^{2+}	$3d^4$	4	4.90	4.80
Mn^{2+}	$3d^5$	5	5.92	5.95
Fe^{2+}	$3d^6$	4	4.90	5.0-5.5
Co^{2+}	$3d^7$	3	3.87	4.4-5.2
Ni^{2+}	$3d^8$	2	2.84	2.9-3.4
Cu^{2+}	$3d^9$	1	1.73	1.4-2.2
Zn^{2+}	$3d^{10}$	0	0	0

Formation of Colored Ions by D Block Elements

Compounds of d block elements have a variety of Colors. When a frequency of light is absorbed, the light transmitted exhibit a Color complementary to the frequency absorbed. Transition element ions can absorb the frequency in the visible region to use it two ways and produce visible Color.

d-d Transition

One way is the excitation of an electron to a higher energy level. In transition element ions, the presence of d-electron and empty d-orbital shall result in Color formation. Valence electron excitation and de-excitation. This is called d-d transition.

D-orbitals are generally degenerate and have the same energy. Presence of ligands that can form coordinate bonds with these ions, remove the degeneracy and split them into two groups namely e_g and t_{2g} d-orbitals. The energy difference (ΔE) depends on the strength of the incoming ligand.

Electrons in the lower d-orbitals can be excited into the higher d-orbitals by absorbing energy in the visible region (λ=400-700nm) and transmit (give) a Color complementary to it.

For example, $[Cu(H_2O)_6]^{2+}$ ions absorb red radiation and appear complimentary blue-green. Hydrated Co^{2+} ions absorb radiation in the blue-green region, and therefore, appear red in the sunlight.

Cupric ion is colorless and in the presence of water molecules becomes blue in color:

- Color of the ions varies with its oxidation state. The Cr^{6+} as in potassium dichromate yellow in color, whereas Cr^{3+} and Cr^{2+} are generally green and blue respectively.
- Color of the compound is dependent on the complexing or coordinating group also. For example, Cu^{2+} shows light blue Color in the presence of water as ligand but the deep blue color in the presence of ammonia as the ligand.
- Transition metal ions which have:
    - Completely filled *d*-orbitals having no vacant d-orbitals for excitation of electrons are colorless. $Cu^{+}(3d^{10})$, $Zn^{2+}(3d^{10})$, $Cd^{2+}(4d^{10})$ $Hg^{2+}(5d^{10})$, and Zn, Cd, Hg are colorless.
    - Transition metal ions which have completely empty *d*-orbitals without d-electrons are also colorless. $Sc^{3++}(3d^{o})$, and $Ti^{4++}(3d^{o})$, ions are colorless.

L-M, M-L and dπ – pπ Bonding

Ligands may donate their p electrons into the empty d orbitals of metal ions. This interaction called as ligand-metal/metal-ligand or dπ – pπ bonding also may give Color to the compounds.

Complex Formation Tendency of D Block Elements

Complex compounds are compounds wherein a number of neutral molecules or anions are bound to a metal. Metals which are a part of the d block elements form many complex compounds owing to their small ionic size, high charge, and relative availability of d orbitals for the formation of bonds.

Transition metal and their ions with their larger nuclear charge and smaller size can attract electrons and Receive lone pair of electrons from anions and neutral molecules into their empty d-orbitals forming coordinate bonding.

Transition elements thus form complex molecules with CO, NO, NH_3, H_2O, F^-, Cl^-, CN^-. Examples of transition metal complexes are, $[Co(NH_3)_6]^{3+}$ $[Cu(NH_3)_4]^{2+}$, $Y(H2O)_6]^{2+}$, $[Fe(CN)_6]^{4-}$, $[FeF_6]^{3-}$, $[Ni(CO)_4]$.

Catalytic Activity of Elements

Catalysts are important for the industrial bulk production of many chemicals. Many d-block elements as metals are in their ionic form are being used as a catalyst on many chemical and biological reactions.

Iron in Haber's process to make ammonia, vanadium pentoxide in the manufacture of sulphuric acid, titanium chloride as Zigler Natta catalyst in polymerization and Palladium chloride in the conversion of ethylene to acetaldehyde are some very important commercial catalytic processes involving d block metals.

Most transition elements act as good catalyst because of,

- The presence of vacant d-orbitals.
- The tendency to exhibit variable oxidation states.
- The tendency to form reaction intermediates with reactants.
- The presence of defects in their crystal lattices.

They take the reaction through a path of low activation energy by,

- Providing a large surface area for absorption and allowing sufficient time to react.
- May interact with the reactants through their empty orbitals.
- May actively interact by redox reaction through their multiple oxidation states.

Alloy Formation in D Block Elements

Atomic radii of the transition elements in any series are not much different from each other. As a result, they can very easily replace each other in the lattice and form solid solutions over an appreciable composition range. Atoms within 15% of the difference in radii can form alloys.

Such solid solutions are called alloys. Alloys are homogeneous solid solutions of two metals or a metal with a non-metal. The alloys of transition metals are hard and high metals are high melting as compared to the host metal.

Various steels are alloys of iron with metals such as chromium, vanadium, molybdenum, tungsten, manganese etc.

Some important alloys are:

- Bronze – Cu(75-90%) + Sn (10-25%);
- Chromium steel – Cr(2-4% of Fe);
- Stainless steel- Cr(12-14% and Ni(2-4%) of Fe;
- Solder- Pb +Sn.

Interstitial Compounds of D Block Elements

Transition metal has a void in their crystal lattice structure. Small non-metallic atoms and molecules like hydrogen, boron, carbon etc. can be trapped in the void during crystal structure formation. These are called interstitial compounds. They are neither ionic nor covalent and non-stoichiometric as in $TiH_{1.7}$, $VH_{0.56}$.

Interstitial compounds have the following properties:

- Their melting points are very high.
- They are extremely hard.
- They have similar conductivity properties when compared to other metals.
- They are unreactive and tend to be chemically inert.

Examples for the interstitial compounds that are formed with transition metals are TiC, Mn_4N, Fe_3H, and TiH_2.

Non-stoichiometric Compounds

Transition metal compounds of different oxidation state may sometimes present together. They may be formed by solid structure defect or by the prevalent conditions. But, this mixture behaves like a single compound.

This compound will not have any finite composition, structure. Non-stoichiometric is shown particularly while combining with group16 (O, S, Se, Te) elements.

Examples: $Fe_{0.94}O$, $Fe_{0.84}O$, $VSe_{0.98}$, $Se_{1.2}$.

Important Compounds of D Block Elements

The D Block Elements form some compounds of vital industrial importance. Some such compounds include:

$K_2Cr_2O_7$ - Potassium Dichromate

This compound is considered very important in the leather industry. It is also used as an oxidant in most azo compound preparing processes.

The structure of the dichromate ion is made up of two tetrahedral that share a single corner with a Chromium-Oxygen-Chromium bond angle of 126°. Potassium dichromate is a strong oxidizing agent. Potassium dichromate is also used as a primary standard in the process of volumetric analysis.

$KMnO_4$ - Potassium Permanganate

The physical appearance of $KMnO_4$ has an intense purple color. It exhibits diamagnetic properties and also weak paramagnetic properties which are reliant on the temperature.

The permanganate ion exhibits diamagnetism due to the absence of unpaired electrons in it. Potassium permanganate is also used as an oxidant in the preparation of various products in organic chemistry. It also finds use in the bleaching of cotton, silk, and wool. It can also be used for the decolonization of oils owing to its strong oxidizing ability.

Variable Oxidation State

Transition elements show variable oxidation states unlike s and p block elements. The oxidation states changes in units of one, e.g. Fe^{2+} and Fe^{+3}, Cu^{+1} and Cu^{+2}.

Scandium can have an oxidation number of (+II) if both s electrons are used for bonding and (+III) when two s and one d electrons are involved. Similarly all the elements show variable oxidation states depending upon the number of electrons available for bonding in their s and d sub-shells.

Transition metals have unique property to show variable oxidation state. This property arises from the fact that the energy levels of 3d, 4d and 5d orbitals are very close to those of 4s, 5s and 6s orbitals respectively and, therefore, electrons from both ns and (n-1)d orbitals can be used, infect they are used, for the formation and bonds by transition metals.

- Minimum oxidation state: All the transition elements except Cr, Cu, Ag, Au and Hg which have a minimum oxidation state of +1 exhibit a minimum oxidation state of +2.
- Maximum oxidation state: Each of the elements in groups III B to VII B can show the maximum oxidation state equal to its group number. For example, Cr in group VIB shows a maximum oxidation state of +6 in $Cr_2O_7^{2-}$ ions.

Most of the elements in VIII group show a maximum oxidation state equal to + 6. However, Ru and Os have a maximum oxidation state equal to +8 which is the highest oxidation state shown by any element.

Relative Stability of Various Oxidation States

- The relative stabilities of various oxidation states of 3d-series elements are due to the extra stability of $3d^0$, $3d^5$ and $3d^{10}$configurations.
- Ti^{4+} ($3d^0$) is more stable than Ti^{3+} ($3d^1$).
- Mn^{2+} ($3d^5$) is more stable than Mn^{4+} ($3d^4$).
- The higher oxidation state of 4d and 5d series elements are generally more stable than those of the elements of 3d series, e.g., Mo, Tc (4d series elements) and W, Re (5d-series elements).
- The highest oxidation states of second and third row elements are encountered in compounds containing the more electronegative elements *viz.* F, O and Cl.

Interstitial Compounds

The compounds formed when small atoms of H, C or N get trapped inside the crystal lattice of metals are known as interstitial compounds. A number of interstitial compounds are formed by the transition metals. Transition metals react with elements such as hydrogen, carbon, nitrogen, boron etc. to form interstitial compounds.

- As vacant spaces of the transition metals are filled up by small atoms, these compounds are hard and rigid.

- The chemical properties of the parent transition metals are not altered during the formation of interstitial compounds.
- However, there are various changes in the physical properties such as density, rigidity, hardness, malleability, ductility, electrical conductivity etc.
- Steel and cast iron are the interstitial compounds of iron which are formed with carbon. In the formation of these compounds, the malleability and ductility of iron are lost to a great extent, but the tenacity of the metal increases.

Transition metals have crystals which are either hexagonal close-packed or face-centered cubic structures.

Both of these lattices are very similar and have two types of holes:

- Firstly, there can be two tetrahedral holes per metal atom that mean there is a hole between four metal atoms.
- Secondly, there can be one octahedral hole per atom that is the hole is between six metal atoms.

Physical and Chemical Properties

Following are the significant physical and chemical properties of the interstitial compounds:

- These compounds have very high melting points, higher than that of the parent transition metals.
- These compounds are very hard. The hardness of some borides is comparable to that of a diamond.
- The conductivity exhibited by them is similar to their parent metal.
- These compounds are chemically inert in nature.

References

- Group-1-elements-alkali-metals, the-s-block-elements, chemistry: toppr.com, Retrieved 4 May, 2019
- General-properties-of-the-group, alkali-metal, science: britannica.com, Retrieved 16 August, 2019
- The-Alkaline-Earth-Metals, Group-2-Elements, 1-s-Block-Elements, Elements-Organized-by-Block, Descriptive-Chemistry, Inorganic-Chemistry: libretexts.org, Retrieved 17 June, 2019
- Physical-properties-of-group-2a-elements-homework-help: tutorsonnet.com, Retrieved 27 January, 2019
- Group-2-elements-alkaline-earth-metals, chemistry: emedicalprep.com, Retrieved 13 August, 2019
- Boron-group-element, science: britannica.com, Retrieved 13 June, 2019
- Boron-family, s-and-p-block-elements: askiitians.com, Retrieved 28 May, 2019
- Group-14-elements: byjus.com, Retrieved 6 February, 2019
- Group-14-elements-carbon-family, the-p-block-elements, chemistry: www.toppr.com, Retrieved 16 July, 2019

- Group-15-elements, the-p-block-elements, chemistry: toppr.com, Retrieved 7 April, 2019
- Nitrogen-group-element, science: britannica.com, Retrieved 12 February, 2019
- Group-16-elements, the-p-block-elements, chemistry: toppr.com, Retrieved 8 July, 2019
- Group-17-elements, the-p-block-elements, chemistry: toppr.com, Retrieved 11 March, 2019
- Complex-formation, general-characteristics-of-transition-elements, d-and-f-block-elements: askiitians.com, Retrieved 9 January, 2019
- Interstitial-compounds, chemistry: byjus.com, Retrieved 29 July, 2019

Permissions

All chapters in this book are published with permission under the Creative Commons Attribution Share Alike License or equivalent. Every chapter published in this book has been scrutinized by our experts. Their significance has been extensively debated. The topics covered herein carry significant information for a comprehensive understanding. They may even be implemented as practical applications or may be referred to as a beginning point for further studies.

We would like to thank the editorial team for lending their expertise to make the book truly unique. They have played a crucial role in the development of this book. Without their invaluable contributions this book wouldn't have been possible. They have made vital efforts to compile up to date information on the varied aspects of this subject to make this book a valuable addition to the collection of many professionals and students.

This book was conceptualized with the vision of imparting up-to-date and integrated information in this field. To ensure the same, a matchless editorial board was set up. Every individual on the board went through rigorous rounds of assessment to prove their worth. After which they invested a large part of their time researching and compiling the most relevant data for our readers.

The editorial board has been involved in producing this book since its inception. They have spent rigorous hours researching and exploring the diverse topics which have resulted in the successful publishing of this book. They have passed on their knowledge of decades through this book. To expedite this challenging task, the publisher supported the team at every step. A small team of assistant editors was also appointed to further simplify the editing procedure and attain best results for the readers.

Apart from the editorial board, the designing team has also invested a significant amount of their time in understanding the subject and creating the most relevant covers. They scrutinized every image to scout for the most suitable representation of the subject and create an appropriate cover for the book.

The publishing team has been an ardent support to the editorial, designing and production team. Their endless efforts to recruit the best for this project, has resulted in the accomplishment of this book. They are a veteran in the field of academics and their pool of knowledge is as vast as their experience in printing. Their expertise and guidance has proved useful at every step. Their uncompromising quality standards have made this book an exceptional effort. Their encouragement from time to time has been an inspiration for everyone.

The publisher and the editorial board hope that this book will prove to be a valuable piece of knowledge for students, practitioners and scholars across the globe.

Index